T0325123

Practical Guide
to ChIP-seq
Data Analysis

Focus Computational Biology Series

This series aims to capture new developments in computational biology and bioinformatics in concise form. It seeks to encourage the rapid and wide dissemination of material for emerging topics and areas that are evolving quickly. The titles included in the series are meant to appeal to students, researchers, and professionals involved in the field. The inclusion of concrete examples and applications, and programming techniques and examples, is highly encouraged.

Systems-Level Understanding of Microbial Communities

Theory and Practice

Aarthi Ravikrishnan and Karthik Raman

Practical Guide to ChIP-seq Data Analysis

Borbala Mifsud, Kathi Zarnack, Anaïs F Bardet

Practical Guide to ChIP-seq Data Analysis

Borbala Mifsud

Kathi Zarnack

Anaïs F Bardet

CRC Press
Taylor & Francis Group
Boca Raton London New York

CRC Press is an imprint of the
Taylor & Francis Group, an **informa** business

A CHAPMAN & HALL BOOK

CRC Press
Taylor & Francis Group
6000 Broken Sound Parkway NW, Suite 300
Boca Raton, FL 33487-2742

© 2019 by Taylor & Francis Group, LLC
CRC Press is an imprint of Taylor & Francis Group, an Informa business

No claim to original U.S. Government works

ISBN 13: 978-1-138-59652-8 (hbk)

Visit the Taylor & Francis Web site at
http://www.taylorandfrancis.com

and the CRC Press Web site at
http://www.crcpress.com

Preface

Over the past decade, the experimental ChIP-seq protocol as well as the associated computational analysis methods evolved rapidly. Researchers face a plethora of possible experimental and analytical ChIP-seq approaches, and the lack of a single perfect recipe can make use of the protocol daunting from the start. Each ChIP-seq experiment needs to be tailored to the protein or modification of interest, to the studied organism and to the biological question. Our aim is to summarise the points that need to be considered when performing such a study in order to obtain high-quality and interpretable data. In this book, we will discuss the importance of experimental and analytical choices and give advice for different scenarios. In particular, we will guide the reader through the computational analysis steps from initial quality control through peak calling to downstream analyses and visualisation. This book will thereby show a full workflow, with alternative paths that are suitable for researchers with diverse bioinformatics experience, using Unix command line and R-based solutions.

We hope that this book will help experimental biologists to design their ChIP-seq experiments with the analysis in mind, and to perform the first analysis steps themselves; moreover, it will support bioinformaticians to understand how the data is generated, what the sources of biases are and which methods are appropriate for the different analysis steps.

Authors

Borbala Mifsud obtained a PhD in molecular biology at the Institute of Molecular Pathology (IMP) in Vienna, Austria. In the laboratory of Thomas Jenuwein, she worked on epigenetic profiling of a histone methyltrasferase mutant mouse, using ChIP-seq. In 2010, she started her postdoctoral work at the EMBL European Bioinformatics Institute (EBI) in the laboratory of Nicholas Luscombe. She is currently an assistant professor at Hamad Bin Khalifa University, Doha, Qatar and honorary lecturer at Queen Mary University London, UK, working on 3D chromatin conformation and the integration of epigenomic data.

Kathi Zarnack earned a PhD in molecular biology at the Max-Planck Institute for Terrestrial Microbiology in Marburg, Germany, working on the impact of posttranslational modifications on transcription factor specificity. Moving into bioinformatics, she then joined the EMBL European Bioinformatics Institute (EBI) in Hinxton, UK, as a postdoctoral researcher in the group of Nicholas Luscombe. Since 2014, she leads a research group on Computational RNA Biology at the Buchmann Institute for Molecular Life Sciences (BMLS), Goethe University Frankfurt, Germany.

Anaïs F. Bardet completed a PhD in computational biology in the laboratory of Alexander Stark at the Institute of Molecular Pathology (IMP) in Vienna, Austria. She studied the conservation of transcription factor binding sites in different *Drosophila* species and developed tools and pipelines for the comparative analysis of ChIP-seq data. She then worked as a postdoctoral researcher in the laboratory of Dirk Schübeler at the Friedrich Miescher Institute for Biomedical Research (FMI) in Basel, Switzerland, where she investigated the sensitivity of transcription factors to DNA methylation. Since 2017, she is a tenured researcher at the National Center for Scientific Research (CNRS) at the University of Strasbourg, France, where she develops projects exploring the regulation of transcription factor binding.

Contents

## CHAPTER 1

Introduction to ChIP-seq

Borbala Mifsud

C HROMATIN immunoprecipitation followed by high-throughput sequencing (ChIP-seq) is amongst the most widely used methods in molecular biology. It aims to determine the genomic sites that interact with a protein of interest. ChIP-seq can be used to identify transcription factor (TF) binding sites or broader patterns of post-translational histone modifications, referred to as histone marks, underlying regulatory elements. This method is therefore essential for deepening our understanding of transcriptional regulation.

1.1 CHIP-SEQ EXPERIMENT

ChIP-seq is an experiment based on antibody immunoprecipitation (IP) performed on a population of cells to define protein binding sites in the genome using high-throughput sequencing (HTS) (Figure 1.1A). Since its first introduction in 2005, HTS has been a rapidly evolving field, resulting in a broad spectrum of available library preparation protocols and sequencing technologies. Standard HTS platforms include Illumina/Solexa, Roche/454, SOLiD (ABI) and Ion Torrent (Life Technologies) as well as single-molecule approaches by Pacific Biosciences and Oxford Nanopore Technologies. The technologies differ in their sequencing concepts, their throughput and run times, the lengths of the obtained sequence information and the error rates, among others (see [1] for review).

Here, we describe the protocol for the Illumina sequencing platform, which is the most widely used. Protocols using other platforms would be similar with slight differences at the library preparation and sequencing steps.

Chromatin preparation The protocol starts with formaldehyde crosslinking of chromatin-bound proteins to DNA, thereby taking a snapshot of how proteins and modifications are distributed along the genome at that point in time. Subsequently, the crosslinked chromatin is sheared until the average DNA fragment length reaches a tight distribution, usually around 200 bp.

> Note: Alternatively, ChIP-seq can be performed without crosslinking, which is mostly referred to as native ChIP. In this case, chromatin is digested by micrococcal nuclease (MNase) to reach the desired DNA fragment size. Full digestion leads to ~147 bp fragments corresponding to the length of DNA wound around a single nucleosome. Native ChIP is often used for histone modifications, but also for other factors when crosslinking would mask the epitope recognised by the antibody. However, it is not advisable to use native ChIP for factors that are loosely bound to chromatin, since their contact with DNA might get lost during the protocol.

Immunoprecipitation In the next step, an antibody specific to the protein or modification of interest is used to recognise the corresponding protein-DNA complexes. These complexes are pulled down and enriched using beads binding to the antibody. After a few washing steps that reduce non-specific binding to the beads, proteins are digested, and the extracted co-purified DNA is subjected to library preparation and HTS.

Library preparation Preparing the sequencing library involves repairing the ends of the DNA fragments, and ligating adapters for sequencing. In case of multiplexing, i.e. running more than one sample on the same lane of the flow cell, an indexed adapter is ligated, which allows assigning the sequenced reads to the multiplexed samples after sequencing. The ligation products are purified and PCR-amplified in order to obtain enough material for sequencing. Amplification being a known source of bias, the number of PCR cycles should be kept to a minimum. The library is then loaded onto the flow cell and sequenced.

Sequencing Sequencing of the sheared DNA fragments is always carried out in a 5'-to-3' direction. The forward and reverse adapters are randomly ligated to either end of the double-stranded DNA fragments. Single-end sequencing starts from either the forward or the reverse adapter, while paired-end sequencing is from both ends. The read length is usually shorter than the co-purified DNA fragments, consequently only the fragment ends are sequenced.

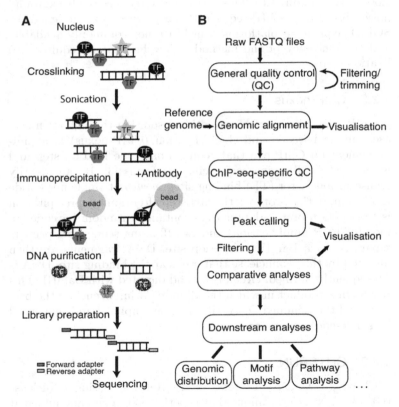

Figure 1.1 (**A**) Schematics of the ChIP-seq experiment. (**B**) ChIP-seq data analysis workflow.

1.2 IMPROVED DETECTION PROTOCOLS

1.2.1 ChIP-exo

ChIP combined with exonuclease digestion (ChIP-exo) achieves almost nucleotide resolution in defining binding sites [2]. After ligation of the first adapter, *E. coli* 5'-3' lamda exonuclease is used to trim the co-purified DNA fragments from the 5' end up to a few nucleotides from the crosslink site. The strands are then separated and ligated to the second adapter for sequencing. This means that the reads represent the remainder of the DNA fragment on the 3' side of the crosslink site, with the start of the read lying at or very close to the binding site. This method also reduces the background noise observed in ChIP-seq. ChIP-exo however was designed for SOLiD sequencing platforms, which are not commonly available, and the efficiency of the method is low because it requires two ligation steps.

1.2.2 ChIP-nexus

ChIP experiments with nucleotide resolution through exonuclease, unique barcode and single ligation (ChIP-nexus) is an improvement on ChIP-exo that requires only one ligation step and is adapted for Illumina sequencing platforms [3]. It can identify transcription factor (TF) binding sites *in vivo* at single nucleotide resolution. In this protocol, the first step after immunoprecipitation is the ligation of an adapter that contains a random barcode (to monitor PCR amplification) and two Illumina sequencing primers separated by a BamHI restriction site. DNA fragments are then digested by the exonuclease that stops at the binding site. DNA is subsequently purified, circularised and digested with BamHI. This results in a product in which the Illumina primers enclose the barcode and the trimmed DNA, that can be amplified and sequenced by single-end sequencing.

1.2.3 CUT&RUN

Cleavage Under Targets and Release Using Nuclease (CUT&RUN) is a technique that is performed *in situ* and aimed at reducing background noise [4]. In this protocol, the unfixed nuclei are immobilised on magnetic beads and incubated with the antibody specific for the protein or modification of interest and protein A-MNase (pA-MN). pA-MN binds to the antibody and cleaves the DNA around the protein that the antibody is bound to. This means that rather than fragmenting all DNA as in standard ChIP approaches, only protein-bound DNA fragments are excised. Then single nucleosomes or protein-DNA complexes are recovered in the supernatant after centrifugation, and the purified DNA is used for library preparation and high-throughput sequencing.

1.2.4 DamID

DNA adenine methyltransferase identification (DamID) works as an alternative to chromatin immunoprecipitation [5]. We wish to mention it here, as this method can work even if there is no specific antibody available for the TF or nuclear protein of interest. DamID is based on fusing the *E. coli* adenine methyltransferase to the protein of interest. This fusion protein then creates N6-methyladenine at GATC motifs. The disadvantages of DamID are (i) that it cannot be used for posttranslational modifications such as histone marks, (ii) that the resolution is determined by the GATC frequency in the genome, and (iii) that it does not directly assess occupancy of protein, but just shows that the protein had been present at a given site. Additionally, DamID requires a specialised analysis pipeline.

1.3 CHIP-SEQ DATA ANALYSIS WORKFLOW

ChIP-seq data analysis comprises several steps (Figure 1.1B). The first step after receiving the raw sequence files is performing standard HTS quality control. The aim is to ensure that the sequencing went well, there was no contamination and the library was complex enough (see Chapter 3). Raw reads are then aligned to the reference genome of the studied organism (see Chapter 4). This can be followed by ChIP-seq-specific quality control, that checks for enrichment in the ChIP-seq sample and excludes over-fragmentation (see Chapter 5). The next step is the central analysis in ChIP-seq, which is identifying the genomic regions that are enriched for the factor of interest, referred to as peak calling (see Chapter 6). Once peak regions are defined, ChIP-seq-specific quality control measures that require peak locations can also be assessed (see Chapter 5). Visualisation of the data should be performed at almost every step (see Chapter 7). It is especially important after peak calling to ensure that the predicted peaks accurately capture the binding pattern. The analyses after peak calling depend on the experimental setup and the biological questions. For instance, if there are replicates and/or more than one condition in the experiment, then the next step could be the comparison between samples (see Chapter 8). Biological replicates will give information about reproducibility and intrinsic biological and technical variability in the data (noise). Differential binding analysis addresses which peak regions show significantly different occupancy between two conditions (i.e. differ more than expected from the variation observed between replicates of the same condition). Finally, peak regions or differential peaks can be used in various downstream analysis steps, such as genomic annotation, gene ontology and pathway analyses, motif discovery or integration with additional genomic datasets (see Chapter 9).

1.4 DESIGNING A CHIP-SEQ EXPERIMENT

1.4.1 ChIP-seq controls

In order to confidently identify enriched regions in a ChIP-seq sample, the read distribution needs to be compared to a background distribution to control for potential biases. The optimal control for a ChIP-seq experiment is to sequence the input DNA purified from the sheared chromatin before the antibody incubation step. Other controls can be prepared using different strategies: A mock IP follows all steps of the ChIP-seq protocol but does not use any antibody. An unspecific IP can be performed by using antibodies against a protein that does not bind to chromatin, e.g. immunoglobulin G (IgG). However, both of these approaches lead to small amounts of co-purified DNA with low complexity, which does not reflect the real background distribution. When studying a histone modification, a pan-H3 or pan-H4 antibody that recognises H3 or H4 irrespective of their modifications is a good alternative, as this captures the underlying nucleosome distribution on which the modifications occur.

1.4.2 Sources of bias

Input samples are necessary to control for a number of biases that can cause over-representation of certain regions. Probably the most important source of bias is the non-uniform fragmentation of the chromatin during sonication or digestion. Compact heterochomatic regions are difficult to shear and become under-represented compared to open euchromatic regions even in input samples. The effect of chromatin compaction is linear, i.e. the more open the chromatin is the higher its representation in the library [6]. Additionally, the non-uniform fragmentation together with a bias due to PCR amplification will lead to over-representation of GC-rich sequences. Consequently, the background distribution will be positively correlated with the GC content of the genome. This is particularly prevalent in mammalian cells, where euchromatic regions are enriched for CpG islands. Therefore, this should be taken into account when comparing CpG-rich regions with others harbouring fewer CpGs.

Another source of bias arises from the computational processing of the data. During the genomic alignment step, only reads mapping unambiguously to unique positions in the genome are retained (see Chapter 4). This will lead to low coverage in repetitive regions. Finally, in cancer samples and cell lines, the genome can considerably differ from the reference genome sequence. Regions that are deleted in the studied cell line will appear as depleted, while duplicated or amplified regions produce more reads and are seemingly enriched.

To account for these biases, it is crucial to always compare a ChIP sample (or at least a set of replicates) to a control sample, which will be used to control for potential false-positive peaks during the peak calling step (see Chapter 6). If the experimental design involves several conditions, the additional comparison of different samples is useful to cancel out potential biases, including for protocols that do not generate input (e.g. ChIP-exo) (see Chapter 8).

1.4.3 Antibody quality

Since ChIP-seq is an antibody-based immunoprecipitation experiment, its efficiency strongly depends on the quality and specificity of the antibody. Selection of a good antibody is therefore of crucial importance. For example, it was previously reported that about a third of commercial ChIP-grade antibodies for histone marks do not work [7]. Additionally, individual antibodies for the same protein might recognise distinct epitopes that can be exposed differently depending on the genomic location (especially in the case of monoclonal antibodies). For example, one antibody specific to a factor might detect promoter occupancy, while another antibody for the same factor might detect occupancy also in intergenic regions. It is therefore advisable to test several antibodies for the same protein of interest and verify their specificity for example by knock-out or knock-down in a Western blot analysis.

1.4.4 Read depth

In order to capture all true binding sites in an experiment, the number of reads sequenced is a determinant factor. The required number of reads depends both on the size of the genome and the binding pattern of the factor of interest (sharp regions for TFs and broad regions for histone marks; see Chapter 6). These two parameters together define the effective genome size, i.e. the number of base pairs (bp) that need to be covered. It also depends on the sensitivity of peak detection that one aims for: in order to identify the most highly enriched peaks (>30x enrichment over input), about a third of the reads are sufficient [8]. Once the sample is sequenced, a saturation analysis can indicate whether all peaks have been detected in the sample for a given number of mapped reads (see Chapter 6.6). In *Drosophila melanogaster*, it has been shown for more than one TF and histone mark that saturation is achieved at around 16 million (M) mapped reads [6]. In contrast, a minimum of 30 M reads are commonly used for TFs in mammals, and more than 60 M reads for broad histone marks. Input samples need to be sequenced at least as deeply as the ChIP samples, as in this case the whole genome needs to be covered.

Note: The required number of reads also depends on the antibody quality and the efficiency of the immunoprecipiation step. Fewer reads are required when higher signal-to-noise ratio is achieved.

Note: Nowadays, sequencing lanes produce a minimum output of 200 to 300 M reads, which means that all ChIP-seq samples can be multiplexed.

1.4.5 Read properties

Decisions related to sequencing reads are the read length and single-end or paired-end sequencing. Single-end and paired-end reads result from sequencing from just one or both adapters that were randomly ligated to either end of the DNA fragments. In most ChIP-seq studies, read length and type of sequencing are not crucial considerations, and the current standard 50-nt single-end sequencing is sufficient for capturing most of the information necessary to infer binding sites. Generally, both longer or paired-end reads have a higher chance of aligning uniquely to the genome, even in slightly repetitive regions. Therefore, a larger proportion of the genome can be covered, and fewer reads get filtered out during alignment [6] (see Chapter 4).

> Note: In case the factor is expected to bind repetitive regions, it is advisable to use the longest possible reads with paired-end sequencing. This will increase the chance for unique mapping if the reads extend beyond the repetitive region. However, repetitive regions remain difficult to study even with longer or paired-end reads and the cost increase it entails might not scale with the expected output.

1.4.6 Replicates

Replicate samples can represent different levels of variability. Technical replicates range from re-sequencing libraries to performing ChIP-seq in replicates from the same cell culture. However, in order to assess biological variability and gain confidence in the identified binding regions, ChIP-seq experiments should always be performed in biological replicates (distinct cell cultures or individuals). In general, a minimum of three replicates is recommended to obtain reasonably reliable results and mandatory for statistical models to hold true, even though many have been adapted to having only two replicates available (see Chapter 8). A bare minimum for each condition would be two replicates for ChIP and one for the corresponding input. More repetitions might be needed if the data is expected to show high intrinsic variability, e.g. when samples are taken from different individuals.

FURTHER READING

Reuter, J.A., Spacek, D.V., and Snyder, M.P. (2015). High-throughput sequencing technologies. *Mol Cell*, 58: 586–597.

Landt S.G. et al. (2012). ChIP-seq guidelines and practices of the ENCODE and modENCODE Consortia. *Genome Res*, 22(9): 1813–1831.

Meyer C.A. and Liu X.S. (2014). Identifying and mitigating bias in next generation sequencing methods for chromatin biology. *Nat Rev Genet*, 15(11): 709–21.

Nakato R. and Shirahige K. (2017). Recent advances in ChIP-seq analysis: from quality management to whole-genome annotation. *Brief Bioinform*, 18(2): 279–290.

Jordán-Pla A. and Visa N. (2018). Considerations on Experimental Design and Data Analysis of Chromatin Immunoprecipitation Experiments. *Methods Mol Biol*, 1689: 9–28.

## CHAPTER 2

Getting Started

Anaïs Bardet

T HROUGHOUT this book, we provide code for processing and analysing ChIP-seq data for transcription factors (TFs) and histone marks. In this chapter, we detail information about the example datasets we use, list the computational requirements and provide basic information on how to run our code.

2.1 CHIP-SEQ DATASETS

Publicly available ChIP-seq datasets often originate from individual publications, which performed in-depth analyses of protein-DNA binding in a specific physiological context. In addition, large-scale consortium efforts like the Encyclopedia of DNA Elements (ENCODE) project collect hundreds and sometimes thousands of datasets. In order to facilitate comparisons and data integration, these are often restricted to selected cell lines and conditions.

High-throughput sequencing data can be retrieved through international data repositories such as:

- NCBI Sequence Read Archive (SRA) accessible through Gene Expression Omnibus (GEO) (`ncbi.nlm.nih.gov/geo`)

- EBI Sequence Read Archive (ERA) accessible through Array Express (`ebi.ac.uk/arrayexpress`)

Data from large-scale consortia are additionally provided on specific websites such as:

- ENCODE (`encodeproject.org`)

- Roadmap Epigenomics (`roadmapepigenomics.org`)

- BLUEPRINT Epigenome (`blueprint-epigenome.eu`) for datasets of haematopoietic cells.

Most of those repositories have some level of redundancy. In this chapter, we explain and provide code to retrieve data from NCBI SRA through GEO, which is one of the most popular repositories for publication-associated data. Datasets are deposited on GEO by authors themselves under a series accession (e.g. GSE67867) including several samples (e.g. GSM1891641). For each sample, raw sequencing reads (e.g. FASTQ files; see Chapter 3) can be found as well as additional already processed files (e.g. mapped reads [BAM files], genomic tracks [bigWig files], list of peaks). While processed data is available for download from GEO directly, raw data is stored on SRA and GEO only links to it. Metadata about the sample can be found by clicking the SRA link (e.g. SRX1280434) at the bottom of the GEO page. It contains information about the sample such as the sequencing instrument (e.g. Illumina HiSeq 2000), layout (e.g. single-end), experimental strategy (e.g. ChIP-seq), material source (e.g. genomic DNA), and experimental protocol as well as the identifiers of the sequencing runs generated for this sample (e.g. SRR2500883). Raw sequencing reads (i.e. FASTQ files) can be retrieved using those identifiers (see Section 2.3).

As examples in this book, we use datasets from Domcke et al., 2015 [9] that have been deposited on GEO under the accession GSE67867. These include ChIP-seq samples for the transcription factor NRF1 and the histone mark H3K27ac (lysine 27 acetylation on histone H3; used as a marker for active enhancers) profiled in either wild type (WT) or DNA methylation mutant (TKO; triple knockout for DNMT1, DNMT3a, DNMT3b) mouse embryonic stem cells. Each sample has been generated in two biological replicates and each condition has a matching input sample. All ChIP-seq libraries were sequenced on an Illumina platform (Illumina HiSeq 2500) using single-end 50-nt reads and are processed using the mm10 mouse reference genome. The samples and their associated metadata are listed in Table 2.1.

The datasets were generated to test the sensitivity of transcription factor binding to DNA methylation, i.e. to test if DNA methylation can block transcription factors from binding [9]. The rationale was that if, in WT cells, some transcription factors cannot bind when the DNA is methylated, new binding sites should appear upon removal of DNA methylation in the TKO cells. By profiling open chromatin regions using DNase-seq, we could iden-

TABLE 2.1   List of datasets and metadata used in this book.

Sample name	GEO ID	SRA ID	Number of raw reads
NRF1_CHIP_WT_1	GSM1891641	SRR2500883	40,570,927
NRF1_CHIP_WT_2	GSM1891642	SRR2500884	40,365,286
H3K27AC_CHIP_WT_1	GSM1891651	SRR2500893	41,972,346
H3K27AC_CHIP_WT_2	GSM1891652	SRR2500894	40,822,025
NRF1_INPUT_WT	GSM1891643	SRR2500885	22,773,779
NRF1_CHIP_TKO_1	GSM1891644	SRR2500886	32,306,980
NRF1_CHIP_TKO_2	GSM1891645	SRR2500887	45,342,909
H3K27AC_CHIP_TKO_1	GSM1891653	SRR2500895	50,829,570
H3K27AC_CHIP_TKO_2	GSM1891654	SRR2500896	45,485,455
NRF1_INPUT_TKO	GSM1891646	SRR2500888	24,937,026

tify new regions that gained transcription factor binding in the
TKO cells compared to WT. A motif analysis identified NRF1 as
a potential candidate to be sensitive to DNA methylation, which
we tested and confirmed by NRF1 ChIP-seq in WT and TKO cells.
In this book, we will guide readers through the analysis of these
NRF1 ChIP-seq datasets.

2.2   COMPUTATIONAL REQUIREMENTS

2.2.1   Computing environment

We provide Unix Bash and R code that generally runs on any
Unix/Linux distribution and should be able to run on a regular
computer (run time is indicated when longer than one minute).
Our example code was run on a Debian GNU/Linux 8.6 distribution on a server with 10 cores, 20 threads, 2.2 GHz processor
and 96 GB of main memory and used about 50 GB of hard-disk
space. When formatted as a PDF file, additional line breaks and
invisible additional characters might have been added to the code.
Therefore, we recommend copying and pasting code directly from
the script available for download online (anaisbardet.cnrs.fr/
practical-guide-to-chip-seq-data-analysis/). All files used
and generated throughout the code are also available as Additional
Online Files.

2.2.2   Data

Additional Online Data Files used in the analyses:

- Blacklisted regions (mm10_blacklist.bed.gz) from (sites.
  google.com/site/anshulkundaje/projects/blacklists)

- Gene annotations generated from ENSEMBL GTF files (ensembl.org) (mm10_genomic_features.bed, mm10_tss.bed)

- List of chromosomes and sizes from UCSC (genome.ucsc.edu) (mm10.chrom.sizes)

- TF motifs from JASPAR 2018 (jaspar.genereg.net)

- ChIPQC and DiffBind sample sheet (NRF1_sample_sheet.csv and NRF1_sample_sheet_with_peaks.csv)

2.2.3 Software

We used the following software and versions:

sratoolkit 2.8.1 (ncbi.nlm.nih.gov/sra/docs/toolkitsoft),

fastqc 0.11.5 (bioinformatics.babraham.ac.uk/projects/fastqc),

cutadapt 1.12 (cutadapt.readthedocs.io/en/stable),

trim_galore 0.4.4 (bioinformatics.babraham.ac.uk/projects/trim_galore),

bowtie2 2.3.0 (bowtie-bio.sourceforge.net/bowtie2),

samtools 1.3.1 (htslib.org),

bedtools 2.27.1 (bedtools.readthedocs.io/en/latest),

peakzilla (github.com/steinmann/peakzilla),

MACS2.1.1 (github.com/taoliu/MACS),

UCSC genome browser (genome.ucsc.edu/),

IGV 2.4.10 (software.broadinstitute.org/software/igv),

IDR 2.0.2 (github.com/nboley/idr),

HOMER 4.8 (homer.ucsd.edu/homer,

MAST 4.11.2 (meme-suite.org/doc/mast.html),

bwtool 1.0 (github.com/CRG-Barcelona/bwtool).

We used the following R and Bioconductor versions and libraries: R 3.3.3 (www.r-project.org) and Bioconductor 3.4 (www.bioconductor.org) with libraries ChIPQC 1.14.0, BiocParallel 1.12.0, NMF 0.21.0, DESeq2 1.20.0, DiffBind 2.6.6, gplots 3.0.1, TxDb.Mmusculus.UCSC.mm10.knownGene 3.4.0, org.Mm.eg.db 3.5.0, ChIPseeker 1.14.2

The code examples in the book require that the software is available in the system path and can hence be executed directly. Otherwise, the commands need to be extended to include full paths. Software manuals can be visualised for Unix commands using `man command` or invoking the software using the option `-h` or `--help`. For a general introduction into Unix commands, please refer to the command-line bootcamp tutorial (`rik.smith-unna.com/command_line_bootcamp`). R command manuals are accessed using `help(command)` or `?command`. A general R introduction is found here: `cran.r-project.org/doc/contrib/Owen-TheRGuide.pdf`.

2.2.4 File formats

BED files offer a flexible way to visualise genomic intervals (mandatory fields: chromosome / start / end; optional fields: name / score / strand) which can be decorated with additional information, such as names and colours. Other commonly used file formats are explained in the subsequent chapters, including FASTQ (Chapter 3), SAM/BAM (Chapter 4), bedGraph/bigWig/bidBed (Chapter 7) and GTF files (Chapter 9). Information about genomic file formats can be found in the FAQ of the UCSC Genome Browser (`genome-euro.ucsc.edu/FAQ/FAQformat.html`).

2.3 DATA RETRIEVAL FROM GEO

SRA developed tools for retrieving datasets from their repository using the command line (SRA Toolkit). The following example code runs `fastq-dump` to retrieve raw sequencing reads in a FASTQ format using the SRA sample ID.

```
# Should be adapted for all datasets in Table 2.1

# Create directory for datasets
mkdir data

# Run fastq-dump (~3 h)
fastq-dump --origfmt --outdir data --gzip -A SRR2500883

# Rename sample according to Table 2.1
mv data/SRR2500883.fastq.gz data/NRF1_CHIP_WT_1.fastq.gz
```

```
# Temporary files (e.g. SRR2500883.sra) are stored in the
tmp directory defined by TMPDIR so make sure you have
enough space and remove them once processed
rm ${TMPDIR}/sra/SRR2500883.sra

# Visualise file
gunzip -c data/NRF1_CHIP_WT_1.fastq.gz | head
```

2.4 CODING TIPS

Folder structure We recommend to establish a folder structure that keeps track of the different analysis steps. All analyses should be run in a base folder (chosen by the user), in which a new subfolder will be created at the beginning of a new analysis step. The example code in this book creates these subfolders and automatically redirects the output files by specifying a path to their specific subfolders. For instance, all alignment files are stored in the folder `reads`. All paths in the example commands are given as relative paths, i.e. starting from the base folder. Our example code uses the following subfolders:

```
<base folder>
  ├── data
  ├── indices
  ├── genomes
  ├── reads
  ├── tracks
  ├── peaks
  ├── changes
  └── motifs
```

Running code for multiple samples Code is generally shown for one sample but can be adapted to run automatically on many samples using a Unix `for` loop:

```
# For one sample
sample=NRF1_CHIP_WT_1
ls data/${sample}.fastq.gz

# For several samples
for sample in NRF1_CHIP_WT_1 NRF1_CHIP_WT_2 NRF1_INPUT_WT
NRF1_CHIP_TKO_1 NRF1_CHIP_TKO_2 NRF1_INPUT_TKO
H3K27AC_CHIP_WT_1 H3K27AC_CHIP_WT_2 H3K27AC_CHIP_TKO_1
H3K27AC_CHIP_TKO_2
do
    ls data/${sample}.fastq.gz
```

```
done
```

Displaying files in Unix   Text files can be visualised directly
in the terminal using the Unix commands `cat`, `head` or `less` commands. Plots can be visualised using the `display` command and
HTML pages opened for example using the `see` or `open` command.

Unix output redirection   Unix code processes files line-by-
line, so the output of a Unix command can be redirected as input
of the next Unix command using pipes `|`. If several redirected in-
puts/files are required, this can be done by using the output of
a command as input file for a another command using `<(...)`.
Some programs accept redirected inputs as files by adding `stdin`
or `-` where the file name is used as an argument.

```
variable="line1\nline2\nline3"
echo -e $variable

# Output redirection as input to the next command using |
echo -e $variable | head -n 1

# Output redirection as input to the current command using
<()
head -n 1 <(echo -e $variable)
```

Working with genomic coordinates   Most of the files generated in this book contain genomic coordinates including chromosome, start and end information (e.g. BED format). Processing such files, that can contain millions of lines (e.g. millions of
mapped reads), is much more efficient when the coordinates are
sorted. Therefore, we recommend always working with sorted files
using the Unix command `sort -k1,1 -k2,2n` that sorts first column 1 by chromosome (character sort) and then column 2 by start
position (numeric sort). The list of chromosomes therefore appears
as chr1, chr10, chr11-19, chr2-9, chrM-X-Y. Many programs will
assume that files are sorted this way and will enable faster processing (e.g. `bedtools -sorted`).

It is also important to keep in mind that genomic coordinates
can be represented as 1-based or 0-based. For example, ENSEMBL
uses a 1-based coordinate system as in GTF or SAM files, which
numbers nucleotides directly using closed intervals i.e. both start
and end coordinates are included. UCSC uses a 0-based coordinate

system as in BED or BAM files, which numbers between the nucleotides and starts at 0. Intervals are therefore given in a half-open format, in which the start position is before the first nucleotide and needs to be removed to convert into 1-based co-ordinates (for more details, see http://genome.ucsc.edu/blog/ the-ucsc-genome-browser-coordinate-counting-systems/). Programs processing genomic coordinates will therefore assume 1-based or 0-based coordinates depending on the type of input file.

Unix command awk   awk is a Unix command for pattern scanning and text processing that we will use throughout our example code. Its syntax can be decomposed in three optional parts:

```
awk 'BEGIN{...} {...} END{...}' file
# Commands to run before reading the file
awk 'BEGIN{...}'
# Commands to run on each line of the file
awk '{...}' file
# Commands to run after reading the file
awk 'END{...}' file
```

For more information, refer to gnu.org/software/gawk/ manual/gawk.html.

Disk space optimisation   Most sequencing datasets represent very big files (>1 GB). In order to optimise disk space usage, it is strongly recommended to always keep them compressed (e.g. .gz using gzip). Those files can then be read without decompressing the original files using the command gunzip -c.

```
# Extracting ouput of a compressed file without
decompressing it
gunzip -c data/NRF1_CHIP_WT_1.fastq.gz | head
```

Analysis reproducibility   It is recommended to keep all commands that were used to run the analysis within a text/script file to ensure its reproducibility. This also enables the deletion of intermediate files that use a lot of disk space and can be recomputed in a reasonable amount of time.

2.5   GRAPHICAL USER INTERFACE TOOLS

The analyses described in this book can also be performed with graphical user interface tools that provide a collection of specialised utilities to cover the whole ChIP-seq data analysis workflow. Two such platforms are the Galaxy web interface [10] and Chipster [11]. Galaxy has several instances with different workflows, e.g. deep-Tools2 [12], which allow the user to perform the quality control, normalisation and visualisation steps, among others.

## CHAPTER 3

General Quality Control

Kathi Zarnack

G ENERAL QUALITY CONTROL (QC) of the high-throughput sequencing (HTS) reads is the first step in every ChIP-seq data
analysis workflow. This chapter introduces quality checks that are
commonly applied to all HTS datasets. These are complemented
by ChIP-seq-specific quality measures in Chapter 5.

3.1 INTRODUCTION

The example ChIP-seq data in this chapter were obtained on
Illumina HiSeq 2500 [9]. Many of the concepts and tools are applicable for HTS data from other sequencing platforms.

3.1.1 FASTQ files

A common format for HTS data is the FASTQ file [13]. Each
entry consists of (i) a unique read identifier (starts with @), (ii) the
determined nucleotide sequence, (iii) a spacer line (starts with #),
and (iv) the quality string with a quality score for each position.
The quality string specifies the base calling accuracy, i.e. the error
probability with which a given nucleotide is identified, encoded in
a `Phred` quality score (Q score): $Q = -10\ log_{10}(error\ probability)$.
In Illumina sequencing, Q scores range between 0–40 (i.e. 40
referring to an error probability of 0.0001). For most applications,
base calls with $Q > 30$, $\in 20$–30 and $Q < 20$ are considered as good,
reasonable and poor quality, respectively. To optimise data storage,

TABLE 3.1   Empirical Q score bins in Illumina FASTQ files.

Q score bin	Representative Q score	Assigned Accuracy
2-9	6	74.88%
10-19	15	96.84%
20-24	22	99.37%
25-29	27	99.80%
30-34	33	99.95%
35-39	37	99.98%
≥40	40	99.99%

the Q scores can be compressed into empirically defined bins which are mapped to one representative value (Table 3.1; for more details, see "Understanding Illumina Quality Scores" at illumina.com).

In the FASTQ file, the Q scores are encoded in ASCII. While earlier versions used an offset of 64 (Phred+64, Illumina 1.3 and 1.4), the more recent versions (Illumina 1.8+) use the original Sanger format of Phred+33. Most tools will automatically determine the offset in the FASTQ file (Figure 3.1A).

3.1.2 Available tools

Several programs offer an easy-to-use quality control for HTS data. Among the most commonly used tools is FastQC which can be run on FASTQ and other file formats from multiple sequencing platforms. Other tools provide additional functionality for data processing, such as the NGS QC Toolkit [14] and the fastx-toolkit (hannonlab.cshl.edu/fastx_toolkit/).

3.2 MEASURES OF HTS DATA QUALITY

3.2.1 Selected quality metrics

The following section focuses on selected quality measures from FastQC which we deem most informative and easy to interpret.

Sequencing quality per position   During many sequencing runs, the base calling accuracy declines with increasing cycles due to progressive degradation of the sequencing chemistry. Moreover, slight quality drops in the first cycles and transient fluctuations during the run can also occur. Such systematic quality changes can be seen in the distribution of Q scores for each sequencing cycle (Figure 3.1A). FastQC issues a warning if 50% of the reads show $Q < 25$ (or 10% with $Q < 10$), indicating that a quality-based trimming of the reads is recommendable (see Section 3.3).

Nucleotide composition per position During the ChIP-seq protocol, the DNA is sheared into random fragments. As a result, the nucleotide composition should be constant throughout the reads (Figure 3.1B). As both strands are sequenced with equal probability, G and C will be equally abundant, but will separate from A and T according to the GC content of the genome.

Average GC content per read In a complex ChIP-seq library, reads are sampled from a large pool of enriched DNA fragments. As a result, the GC content per read is expected to follow a normal distribution (Figure 3.1C,D). Deviations, such as needles or shoulders, indicate a biased subset of reads or a contamination from an organism with different GC content (Figure 3.1C). Some of these biases resolve when filtering against adapters (Figure 3.1D; see Section 3.3).

> Note: The enriched DNA fragments reflect the sequence specificity of the studied transcription factor (TF) or histone mark. In some cases, the GC content may therefore deviate from the uniform distribution e.g. if a TF binds in CpG islands or low-complexity regions such as telomers.

Other quality metrics Several metrics address the library preparation quality. In a complex ChIP-seq library, the majority of enriched fragments will be unique. A high level of duplication, i.e. identical HTS reads, indicates enrichment biases, such as PCR over-amplification due to insufficient amounts of starting material. An excess of free primers indicates that these were not efficiently removed during library preparation (see Chapter 1.1).

In addition, the adapters that were used for sequencing will appear in the reads if the DNA fragments are shorter than the read length. With an average DNA fragment length of 200 bp, adapters should be largely absent in short reads, but start appearing around position 70-100 bp in longer reads. A high adapter content during early sequencing cycles indicates an over-fragmentation of the library. `FastQC` will by default search for several commonly used Illumina adapters. Custom adapters can be provided by the user.

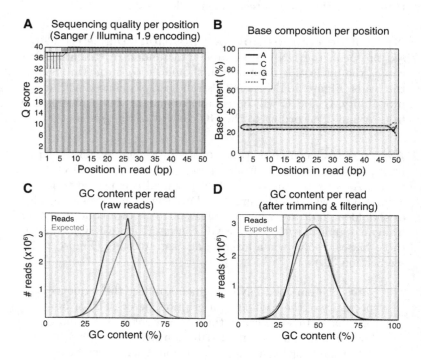

Figure 3.1 **General HTS quality metrics.** (**A**) Distribution of Q scores for each sequencing cycle. Whiskers represent 10 and 90 percentiles. Background shading indicates good, reasonable and poor quality base calls. (**B**) Nucleotide composition per sequencing cycle. Line charts representing the relative abundance of A, C, G and T. (**C,D**) GC content per read (black) across all reads in the library before (C) and after (D) trimming and filtering. Expected normal distribution (grey) estimated from modal GC content as a proxy of the GC content of the underlying genome. Analyses for sample NRF1_CHIP_WT_1 and NRF1_CHIP_WT_1_trimmed modified from FastQC. Images were kindly provided by Cornelia Rücklé and Mirko Brüggemann.

3.2.2 FastQC

The following code shows how to run FastQC from the command line using default settings. The software also offers a graphical user interface. The html output can be visualised in any web browser.

```
sample=NRF1_CHIP_WT_1

# Run FastQC (~3 min)
fastqc -q -o data data/${sample}.fastq.gz

# Visualise html output file
see data/${sample}_fastqc.html
```

3.3 TRIMMING AND FILTERING

Next, the HTS reads are cleaned from excess adapter sequences and low-quality bases. This can be achieved by completely removing the respective reads, thereby maintaining equal read lengths throughout the dataset, or by trimming them.

Note: For comparison of samples, it is important that the samples have the same read length. Different read lengths mean that the mappable proportion of the genome differs, which could lead to artefacts in the comparative analysis.

3.3.1 Adapter removal

If enriched DNA fragments are shorter than the read length, the HTS reads will extend into the downstream adapter. Since the included adapter sequences will impair genomic alignment, they need to be removed prior to genomic alignment.

The stringency of adapter matching relies on several parameters, including the minimum required overlap and the maximum mismatches. A low stringency at this step ensures that most adapters will be detected. Different matching modes specify allowed adapter positions within the read and which part to remove.

Note: Since the ChIP-exo and ChIP-nexus protocols produce DNA fragments that are considerably shorter than 200 bp, the fraction of HTS reads with trailing adapter sequences is higher than for standard ChIP-seq data. Adapter removal is therefore of particular importance.

3.3.2 Low-quality trimming

Low-quality sequences in a read may impair its genomic alignment. In our experience, most ChIP-seq datasets with short read length will not require low-quality trimming. However, if substantial quality drops are visible (see Section 3.2), the reads should be either trimmed or discarded.

A common approach to remove low-quality sequences is to trim each read at the first position where the Q score falls below a given threshold (often $Q < 20$). Alternatively, individual low-quality positions can be tolerated, such that the read is trimmed only if the average Q score in a sliding window falls below a given threshold.

As another option, all reads in the dataset can be cut at a point before low-quality sequences begin to prevail. This avoids read length differences and allows integration of ChIP-seq datasets with different sequencing lengths (e.g. to compare samples with 50 nt and 100 nt reads, all reads should be trimmed to 50 nt).

Note: Analysis of allele-specific binding on the maternal or paternal chromosome requires reliably calling single-nucleotide variants (SNVs). For this, a higher Q score threshold should be used (often $Q > 30$), and window-based trimming is not advisable.

Performance check We recommend to always re-run `FastQC` on the trimmed datasets. In addition, most tools return the number of trimmed reads, removed nucleotides and removed reads. If strong differences appear between datasets, the quality of the samples should be revisited. If necessary, the trimming parameters should be adjusted to keep the samples as comparable as possible.

3.3.3 Trim Galore!

Popular tools for quality trimming and filtering are Trim Galore! (see below), FLEXBAR [15] and Trimmomatic [16].

```
# Run Trim Galore! (optional ~10 min)
trim_galore -q 20 --stringency 2 -o data
data/${sample}.fastq.gz

# List output file
ls data/${sample}_trimmed.fq.gz

# Re-run FastQC to check the trimming performance
```

## CHAPTER 4

Genomic Alignment

Kathi Zarnack

A LIGNMENT of the high-throughput sequencing (HTS) reads to a reference genome identifies the origin of the co-purified DNA fragments. This chapter introduces different alignment concepts and considerations for parameter settings as well as measures to assess mapping quality.

4.1 INTRODUCTION

The goal in genomic alignment is to find the most likely origin of a HTS read in the reference genome. In addition to the large genome size and the immense quantities of reads, the task is further complicated by possible mismatches between the read and the reference sequence. These sequence deviations can arise from amplification or sequencing errors in the HTS reads, but also from genomic variation or errors at the level of the reference genome.

4.1.1 Alignment concepts

Since ChIP-seq reads are directly derived from DNA fragments, the data is usually aligned with "contiguous short read mappers". Many of these alignment algorithms employ the "seed-and-extend" approach. In the first step, the algorithm identifies k-mer seeds, i.e. segments of the read of defined length that precisely map to a given location in the genome [17]. Depending on the algorithm, seed matches have to be exact or can tolerate a given number of mismatches. In the second step, the seeds are extended in both directions using dynamic programming to reach the maximum mappable length without gaps, and finally generate the full alignment.

Based on the conceptual differences, available algorithms differ in terms of mapping accuracy (sensitivity and precision) as well as computational performance (run time and memory). Differences are also influenced by the choice of seed length, whereby shorter seeds result in increased sensitivity, whereas longer seeds allow faster searches.

Most algorithms assign a quality score which estimates the accuracy of the obtained alignment (see Section 4.2.4). In some cases, this score takes into account the base calling accuracy of the read (i.e. Q score in the FASTQ file) to weigh mismatches.

4.1.2 Available tools

A broad range of alignment tools are available that differ in terms of concept, indexing method, computational performance and mapping accuracy (for a more detailed review, see [17]). A very popular tool for ChIP-seq data is the "seed-and-extend"-based algorithm Bowtie 2 which offers high accuracy and speed [18]. Due to its efficient index encoding, Bowtie 2 has relatively small memory requirements, supporting its application on a common laptop or desktop computer. In addition, many specialised applications were designed for specific use cases. For instance, Bowtie 2 explicitly supports the alignment of reads from emerging third-generation sequencing approaches. Alternatively, the ENCODE project relies on the algorithm BWA to align hundreds of ChIP-seq datasets in an efficient and reproducible manner [19].

4.2 PARAMETERS AND CONSIDERATIONS

4.2.1 Mismatches

Due to sequencing errors but also to single nucleotide variants, some reads will not perfectly match the reference genome. To avoid losing these reads, a given number of mismatches should be allowed during alignment. The optimal threshold depends on the type of sample and experimental setup. Most alignment algorithms allow to specify either an absolute number of mismatches per alignment, or a mismatch frequency relative to read length (for parameter settings in Bowtie 2, see Section 4.3).

Note: A higher number of mismatches should be allowed for ChIP-seq experiments from highly mutagenised cells, such as cancer cells, or when aligning to a low-quality reference genome. In addition, some platforms show considerably higher error rates than others. For instance, Illumina sequencing usually introduces less than 0.1%, whereas error rates can go above 10% for third-generation sequencing approaches [20].

4.2.2 Multi-mapping

Multi-mapping, i.e. reads that align equally well to multiple locations in the genome, presents a considerable challenge in short read alignment [21]. The most common sources for such ambiguous alignments are repetitive regions, such as *Alu* elements which make up more than 10% of the human genome. Repetitive regions can also originate from segmental or even whole-genome duplication, e.g. in *Arabidopsis thaliana*.

Different concepts have been introduced to deal with multi-mapping events. Following a conservative approach, many workflows keep only uniquely mapping reads for further analysis. Alternatively, multi-mapping events can be taken into account by using all or just one randomly selected alignment position.

Note: When multiple alignment positions of a read are reported, the number of alignments can be considerably higher than the total number of reads.

Since the co-purified DNA fragments in ChIP-seq are around 200 bp, binding sites within shorter repetitive regions will still be captured if a sufficient number of reads uniquely aligns around the repeat. If preliminary analysis indicates binding to a certain type of repeat, alignment can also be performed against the consensus repeat sequence extracted from Repbase (`girinst.org/repbase/`). This usually achieves higher coverage and allows for a more accurate quantification of repeat binding. Additional considerations when studying proteins that are expected to bind in repetitive regions, can be found in Chapter 1.

4.2.3 Other parameters

Genome version   Most reference genomes exist in multiple versions. It is usually advisable to use the most recent release. For most analyses, it is sufficient to take into account the conventional chromosomes (e.g. chr1-22,X,Y,M in human) and to remove any alternative scaffolds, as these may introduce alignment ambiguities.

Note: If no reference genome sequence is available for the studied organism, the ChIP-seq data can be aligned to the genome of a closely related species. In this case, allowing a higher number of mismatches is advisable to account for differences in the genome sequence. Alternatively, it is possible to reconstruct binding sites by *de novo* assembly of the ChIP-seq reads [22]. Although the location of the binding site within the genome remains unknown, this approach allows for certain downstream analyses, such as motif discovery. In addition, the increased length of the assembled binding sites compared to individual reads can facilitate alignment to a closely related genome sequence. Combining reads from both ChIP and input samples may allow for assembly of larger contigs.

Note: In the analysis of allele-specific binding, heterozygous single nucleotide variants (SNVs) present in the reads are used to discriminate binding events on the maternal or paternal copy of a given chromosome. Generally, this can be assessed by overlaying the SNV information with mismatches in the read alignments. One may consider to adjust the number of allowed mismatches to capture all read variants. However, in order to avoid reference biases, a recent study directly aligned the ChIP-seq reads to the haploid parental chromosomes which were reconstructed from SNV information [23].

Soft-clipping Some alignment algorithms increase mapping rates by so-called soft-clipping, i.e. nucleotides at either end of a read can be excluded from the alignment. This can be useful to get around low-quality sequences at read ends, but also to position reads that overlap with unannotated genomic rearrangements.

Note: No soft-clipping at the 5' end should be allowed for single-nucleotide resolution techniques such as ChIP-exo or ChIP-nexus (see Chapter 1.2).

Single-end versus paired-end reads As explained in more detail in Chapter 1.4, most ChIP-seq experiments use single-end sequencing. If paired-end reads are available, both mates should be aligned together to improve alignment accuracy. This results in an increased fraction of uniquely mapped pairs compared to single-end sequencing.

4.2.4 Output format

Read alignments are commonly reported in a BGZF-compressed BAM file or the corresponding human-readable SAM counterpart [24]. The header section provides details about the initial FASTQ file, the applied alignment software (including the chosen parameters) and the used reference genome. In the alignment section, each alignment is described by the read's name, sequence and quality string together with information on the alignment coordinates in the reference genome. An extended CIGAR string describes matching read fractions, insertion, deletions, etc. Optional fields allow to add additional tags, which can e.g. report on the number of alignments of a given read in the genome (multi-mapping) or duplication.

SAMtools is a software package that provides various utilities to process SAM/BAM files [24]. In particular, sorting and indexing allows fast retrieval of alignments overlapping a particular genomic region, without loading all alignments into memory. Similarly, Picard (broadinstitute.github.io/picard/) offers a collection of command line tools including some options for advanced filtering of alignments.

4.3 GENOMIC ALIGNMENT WITH BOWTIE 2

This section demonstrates how to align ChIP-seq data with Bowtie 2, followed by several post-processing steps. First, the reference genome sequence (here, mouse genome, release mm10) is downloaded from UCSC and converted into a Bowtie 2 index. Bowtie 2 is then run on an example FASTQ file, and the resulting BAM file is filtered, sorted and indexed.

Bowtie 2 is a very fast and memory-efficient alignment algorithm [18]. In contrast to the previous version, Bowtie 2 no longer offers parameters to explicitly define thresholds on mismatches and multi-mapping. Instead, Bowtie 2 implements a scoring scheme which assigns user-configurable penalties to gap opening and extension etc. Mismatch penalties are weighted by the Q score of the respective nucleotide in the FASTQ file (see Chapter 3.1). An alignment is considered as valid if the total alignment score lies above a user-defined threshold. The alignment score is reported in the MAPQ field of the SAM/BAM file. For more details, please refer to the Scoring Options section in the Bowtie 2 manual (bowtie-bio. sourceforge.net/bowtie2/manual.shtml#scoring-options).

Alignments generated by Bowtie 2 are usually post-processed by filtering for a given alignment score. A commonly applied filter is MAPQ > 10 which generally keeps uniquely mapping reads only. Filtering can be performed with SAMtools as shown below. BED files (see Chapter 2.2.4) can be easily generated from BAM files using the bamToBed utility of bedtools.

```
# Create directories
mkdir reads
mkdir -p genomes/mm10
mkdir -p indices/mm10


# Genome sequence can be downloaded from UCSC
# e.g. for mouse genome (mm10)
wget http://hgdownload.soe.ucsc.edu/goldenPath/mm10/bigZips/
chromFa.tar.gz -O genomes/mm10/chromFa.tar.gz
tar -zxvf genomes/mm10/chromFa.tar.gz -C genomes/mm10
# Keep only the conventional chromosomes (chr1-19,X,Y,M)
rm genomes/mm10/chromFa.tar.gz genomes/mm10/*random*
genomes/mm10/chrUn*


# Generate Bowtie 2 index
cd genomes/mm10/
bowtie2-build
chr1.fa,chr10.fa,chr11.fa,chr12.fa,chr13.fa,chr14.fa,chr15.
fa,chr16.fa,chr17.fa,chr18.fa,chr19.fa,chr2.fa,chr3.fa,
chr4.fa,chr5.fa,chr6.fa,chr7.fa,chr8.fa,chr9.fa,chrM.fa,
chrX.fa,chrY.fa ../../indices/mm10/mm10
cd ../..


sample=NRF1_CHIP_WT_1


# Run Bowtie 2 (~1 h)
# Input file can be .fastq.gz or _trimmed.fq.gz if trimming
was performed
# Option -p n if n cores are available
bowtie2 -x indices/mm10/mm10 -U data/${sample}.fastq.gz >
reads/${sample}.sam
```

```
# Filter for -q 10 (~4 min)
samtools view -Sb -q 10 reads/${sample}.sam >
reads/${sample}_nonSorted.bam

# Sort BAM file by genomic coordinates (~5 min)
samtools sort reads/${sample}_nonSorted.bam >
reads/${sample}.bam

# Remove intermediate files
rm reads/${sample}.sam reads/${sample}_nonSorted.bam

# One-line alternative (no intermediate files)
bowtie2 -x indices/mm10/mm10 -U data/${sample}.fastq.gz |
samtools view -Sb -q 10 | samtools sort >
reads/${sample}.bam

# Index BAM file to allow fast access
samtools index reads/${sample}.bam

# Visualise file in SAM or BED format
samtools view reads/${sample}.bam | head
bamToBed -i reads/${sample}.bam | head
```

Once the BAM files have been sorted and indexed, basic alignment metrics allow assessment of the alignment performance.

- The fraction of aligned reads, i.e. reads from the initial FASTQ file that could be successfully aligned to the reference genome, should be as high as possible, usually >70%. Libraries with alignment rates <50% may contain major contamination or other flaws and should be excluded.

- The fraction of unique read start positions in the BAM file serves as a measure of library complexity. In the input sample, reads should be spread across the entire genome, while in the ChIP sample, they should accumulate in certain locations. As a consequence, the fraction of unique read starts should always be lower in the ChIP sample compared to the input. In addition, this fraction tends to decrease for TFs compared to histone marks, reflecting their more selective binding at distinct sites. In our experience, ChIP-seq samples commonly show around 80% unique read start positions (compared to around 15% in RNA-seq).

- The relative abundance of the most frequent individual reads allows us to check for potential artefacts. High numbers of such duplicates can compromise library complexity.

The following code examples were implemented using awk (see Chapter 2.4).

```
# Number of raw reads (i.e. every 4th line of FASTQ file)
gunzip -c data/${sample}.fastq.gz | awk 'END{print NR/4}'
gunzip -c data/${sample}.fastq.gz | awk '(NR%4==2)' | wc -l

# Number of aligned reads
bamToBed -i reads/${sample}.bam | wc -l

# Number of unique positions chromosome + start
bamToBed -i reads/${sample}.bam | awk
'{if($6=="+"){position=$1":"$2}else
if($6=="-"){position=$1":"$3};total[position]=1}END{print
length(total)}'

# One-line command
# sample / raw_reads / mapped_reads / percent_mapped /
unique_positions / percent_unique / most_repeated_read /
number_repeated / percent_repeated
bamToBed -i reads/${sample}.bam | awk -v OFS="\t" -v
sample=$sample -v raw=$(gunzip -c data/${sample}.fastq.gz |
awk -v OFS="\t" '(NR%4==2)' | wc -l)
'BEGIN{max=0}{total++;if($6=="+"){position=$1":"$2}else
if($6=="-"){position=$1":"$3};count[position]++;
if(count[position]>max){max=count[position];maxPos=position}}
END{totalPos=length(count); print
sample,raw,total,total*100/raw,totalPos,totalPos*100/total,
maxPos,count[maxPos],count[maxPos]*100/total}'
```

FURTHER READING

Ye H. et al. (2015). Alignment of Short Reads: A Crucial Step for Application of Next-Generation Sequencing Data in Precision Medicine. *Pharmaceutics* 7(4):523–541.

## CHAPTER 5

ChIP-seq-specific Quality Control

Borbala Mifsud

C HIP-SEQ DATA QUALITY is usually examined by specific metrics after genomic alignment in addition to the quality control of raw sequencing reads (see Chapter 3). In this chapter, we will focus on the R/Bioconductor package ChIPQC that provides a comprehensive ChIP-seq-specific quality control analysis [25]. Other methods are listed in the section Further Reading at the end of the chapter.

Note: A few of the following metrics also require identified peak regions (see Chapter 6).

5.1 CHIP-SEQ-SPECIFIC QUALITY METRICS

5.1.1 Signal enrichment

ChIP-seq enriches for regions that are bound by the transcription factor (TF) or modified histone of interest. Therefore, one of the most important quality control steps is to check for the presence and types of enriched regions. A straightforward score that reflects enrichment is the Standardised Standard Deviation (SSD) of the coverage developed in the R/Bioconductor package htSeqTools [26]. A high SSD score means that there are regions with high signal in the sample. This measure, however, does not assess the structure of the signal and can be strongly affected by artefacts.

Enrichment in peak regions A simple way to check for enrichment is to call peaks (see Chapter 6) in the sample and determine the fraction of reads in peaks (FRIP). In a good ChIP-seq library, FRIP should be >5% for a TF and >25-30% for broad histone modifications or RNA polymerase II (Pol II) (Figure 5.1A).

Enrichment in expected genomic regions Most factors will be enriched in one or a few of the following annotated genomic regions: promoters, exons, introns, 5' or 3' untranslated regions (UTRs). This allows for assessing the immunoprecipitation (IP) efficiency in the library before peak calling. The relative enrichment of ChIP-seq reads within these regions compared to the features' genomic proportion indicates whether the expected genomic distribution was captured. For instance, the NRF1 ChIP-seq samples show high enrichment in promoters, in particular near the transcription start site (TSS), and in 5' UTRs (Figure 5.1B).

Read depth distribution Enrichment is reflected in the number of base pairs with increased read depths, i.e. covered by a higher number of reads. Input samples generally show many positions that are covered with only few reads, and the number of positions with higher coverage drops drastically. Good ChIP-seq samples will also have many positions with a low read count, representing the background, but these samples will show an increased number of highly covered positions as part of peak regions. This can be visualised by plotting the number of positions that are covered at a given read depth. When input and ChIP-seq samples are compared in the same plot, the two lines should diverge towards the higher depths, as seen for the NRF1 samples (Figure 5.1C).

Enrichment in blacklisted regions For many organisms and cell lines, blacklists are available that collect regions with commonly observed ChIP-seq enrichment artefacts, often located at telomeres and centromeres (see Chapter 2.2.2). While these regions only represent about 0.5% of the genome, they can in some cases capture >10% of the ChIP-seq signal. Importantly, the blacklists allow us to directly identify and remove peaks that originate from such artefacts. ChIPQC calculates the fraction of reads in blacklists (FRIBL), which is very low (<0.05%) in the NRF1 samples. If a sample contains a high proportion of such artefactual reads, they should be filtered out before peak calling, as they can affect the

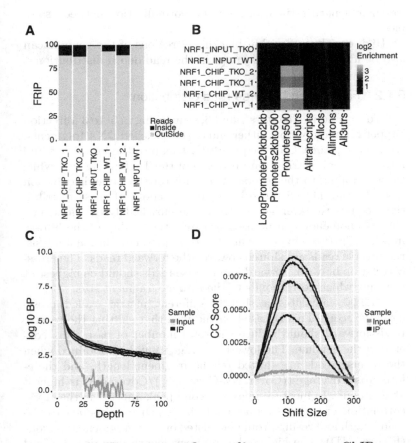

Figure 5.1 **ChIP-seq-specific quality measures.** ChIP-seq-specific quality control was performed for NRF1 ChIP-seq samples in wild type and DNMT triple knockout cells, and for the corresponding input samples. (**A**) Fraction of reads in peaks (FRIP) shows the percentage of reads that maps to the merged peak set. (**B**) Relative enrichment in annotated genomic regions (mm10). (**C**) Read depth distribution shows the $\log_{10}$-transformed number of base pairs that are covered by a given number of reads (Depth). (**D**) Cross-coverage plot shows the CC Score=(Coverage$_0$-Coverage$_n$)/Coverage$_0$ at the given reverse read shift (bp). Images modified from ChIPQC.

fragment length estimation and the normalisation between samples.

High enrichment within blacklisted regions of the genome can also be identified by sharp jumps in the read depth distribution.

5.1.2 Forward and reverse read distribution

As discussed in Chapters 1 and 6, sequencing in 5' to 3' direction happens randomly from either end of the enriched DNA fragments. As a consequence, ChIP-seq reads that map to the forward strand form a normal distribution upstream of the TF binding site, while those mapping to the reverse strand mirror the same distribution downstream. The distance between the modes of these distributions equals the average length of the enriched DNA fragments. This phenomenon can be used to assess the quality of the library in a TF ChIP-seq experiment [27]. In a cross-correlation analysis, the reverse reads are shifted towards the forward reads. The cross-correlation between forward and reverse reads should be highest at the shift which corresponds to the average fragment size.

ChIPQC calculates a slightly different metric, called cross-coverage score. This is based on the idea that shifting the reverse reads forward should reduce the total number of base pairs covered in the genome. The lowest coverage should be obtained when the reverse reads are shifted by the fragment length, and therefore cross-coverage $((\text{Coverage}_0\text{-Coverage}_n)/\text{Coverage}_0)$ is highest at this point. In most samples, a second peak appears in the cross-correlation or cross-coverage analysis, which corresponds to the read length and results from blacklisted or over-fragmented regions (Figure 5.1D). The higher the fragment-length peak is compared to the read-length peak, the better the library quality. ChIPQC provides a cross-coverage score at fragment length, at read length and the proportion of the two (Rel_CC). In our example of the NRF1 samples, the highest cross-coverage peaks are at >100 bp, and the proportion between fragment- and read-length peaks is >1.5, indicative of high-quality samples.

5.1.3 Duplicate reads

It is also important to check for the level of duplication within the sample. Duplication level alone, however does not give an indication of sample quality. Duplication can be a sign of low quality ChIP-seq samples, as duplicated reads can come from PCR over-amplification of libraries with low starting material and therefore low complexity, as well as from blacklisted regions, which are arte-facts and should be removed. However, deeply sequenced, good quality ChIP-seq samples will have high level of duplication due to the enrichment of a relatively small fraction of the genome. If the library was complex enough (see Chapter 6), then removal of duplicates would lead to removal of real signal and enrichment would be saturated, which can mask differences between samples. If blacklisted regions are filtered and library complexity is high, duplicated reads should be kept in the analysis.

5.2 CHIPQC

The R/Bioconductor package `ChIPQC` requires that cer-tain information about your sample is organised in a table, which is provided either as a `data.frame` or a `.csv` file (see NRF1_sample_sheet_without_peaks.csv in Additional Online Data Files at `anaisbardet.cnrs.fr/` `practical-guide-to-chip-seq-data-analysis/` for descrip-tion of the NRF1 dataset). After peak calling, the peaks can be included in the sheet to calculate peak-related quality metrics (see NRF1_sample_sheet.csv).

`ChIPQC` analysis can be performed in R using the following com-mands.

```
# Open R
R

# Load libraries and set up the environment
library(ChIPQC)
library(BiocParallel)
register(SerialParam())

# Run analysis and create report of results (~10 min)
qc=ChIPQC("NRF1_sample_sheet_without_peaks.csv", "mm10",
blacklist="mm10_blacklist.bed", consensus=TRUE, bCount=TRUE)
ChIPQCreport(qc, facet=FALSE, colourBy="Condition")

# Close R
q()
```

FURTHER READING

Carroll T.S. et al. (2014). Impact of artifact removal on ChIP quality metrics in ChIP-seq and ChIP-exo data. *Front Genet*, 10;5: 75.

Diaz A. et al. (2012). CHANCE: comprehensive software for quality control and validation of ChIP-seq data. *Genome Biol*, 13(10): R98.

Ramírez, F. et al. (2016). deepTools2: A next Generation Web Server for Deep-Sequencing Data Analysis. *Nucleic Acids Res*, 44(W1): W160–165.

## CHAPTER 6

Peak Calling

Anaïs Bardet

P EAK CALLING is the core of each ChIP-seq data analysis
pipeline as it identifies the genomic regions occupied by the
transcription factor (TF) or histone mark of interest. This chapter
goes through the different types of ChIP-seq signals and how these
affect the identification of peaks. It describes the general strategy employed by most peak callers, and introduces several existing
tools and their specific features. Finally, it provides example code
for peak calling on ChIP-seq data for a TF and a histone mark.

6.1 CHIP-SEQ SIGNAL TYPES

The goal of a ChIP-seq experiment is to identify genomic regions
that interact with a TF or histone mark of interest. These binding sites appear as regions with high read densities, referred to as
peaks. As introduced in Chapters 1 and 5, high-throughput sequencing of ChIP samples is performed randomly from either end
and does not cover the complete length of the enriched DNA fragments. As a consequence, reads mapping to the forward and reverse
strand form a characteristic bimodal distribution (Figure 6.1). Depending on the type of the protein under investigation, the shape
of the ChIP-seq signal and therefore also the peak calling strategy
differ slightly.

6.1.1 Sharp signal for transcription factors

TFs often recognise specific DNA sequence motifs. Therefore,
the enriched DNA fragments are centred around the motif, leading
to sharp "peaky" regions of enrichment (Figure 6.1A). TFs that

show homotypic binding are characterised by clusters of multiple binding sites in close proximity that will appear as broader regions merging two or more specific peaks.

6.1.2 Broad signal for histone marks

Histone marks generally span several nucleosomes that are not specifically positioned on the DNA sequence but rather depend on the position of neighbouring TFs. Therefore, DNA fragments covering the same region correspond to several nucleosomes that can be loosely positioned on the DNA. As a result, the ChIP-seq signal appears as broad regions of enrichment that can reach several kilobases (Figure 6.1B).

6.1.3 Mixed signal for RNA polymerase II

The localisation of RNA polymerase II (Pol II) is used as a proxy for gene transcription. In some cases, Pol II is paused at gene promoters indicative of regulation at the level of transcription initiation. The ChIP-seq signal can therefore appear as a mix of sharp signals at promoters, corresponding to initiation or pausing, and broad signals within gene bodies, corresponding to transcription elongation (Figure 6.1C).

6.2 GENERAL PEAK CALLING STRATEGY

Many peak callers follow the same general framework, which slides a window along the genome, calculates an enrichment of reads in ChIP versus input samples and defines a significance score corrected for multiple testing.

6.2.1 Estimation of fragment size

Since the ChIP-seq experimental protocol includes a step of size selection, the DNA fragments have a tight size distribution, usually around 200 bp, which represents the resolution of the data. The original fragment size of reads resulting from a single-end sequencing strategy can then be estimated by calculating the distance between the forward and reverse strand distributions within the most enriched regions in the genome (Figure 6.1). When paired-end sequencing is applied and fragments can be reconstituted from the reads, the average fragment size is used.

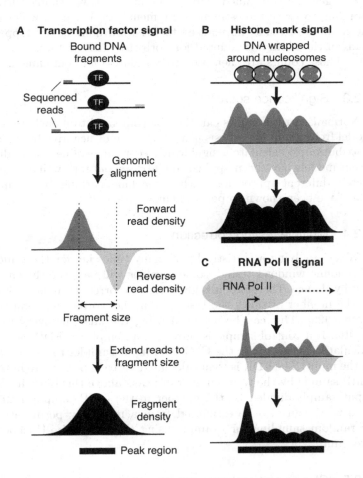

Figure 6.1 **Characteristic ChIP-seq signals for different types of studied proteins.** Experimental design, read density, fragment density and peak regions for (**A**) sharp signal for transcription factors, (**B**) broad signal for histone marks, and (**C**) mixed signal for RNA polymerase II.

6.2.2 Enrichment of reads

The genome is scanned using a sliding window, usually corresponding to twice the estimated fragment size. For each window, the number of reads, normalised to the total number of mapped reads in the library, is counted for both the ChIP and input samples. Those counts are then used to calculate a fold enrichment.

6.2.3 Significance score

Normalised read counts can then be compared to a background model from a null hypothesis using a Poisson or negative binomial distribution to calculate a significance score or p-value. Many different models have been applied to ChIP-seq data, as well as completely different approaches, such as machine learning, but simple ones have been shown to perform equally well [28].

6.2.4 Multiple testing correction

When applying a statistical test many times i.e. for thousands of genomic windows tested, some p-values will pass the threshold just by chance. Therefore, it is important to correct them according to the number of times the test was run i.e. correct for multiple testing [29]. This can be achieved using the false discovery rate (FDR). If a control sample is provided, an empirical FDR can be calculated by swapping the ChIP and input samples to call peaks in the input. The FDR is then calculated for each peak score in the ChIP sample by the total number of peaks above this score in the input sample divided by the number in the ChIP sample. FDRs or q-value can also be estimated from a model by permutation or random sampling for example using the Benjamini-Hochberg procedure [29].

6.2.5 Choice of thresholds

The number of peaks found by different peak calling methods is highly dependent on the thresholds and parameters used, and should therefore be considered with caution. What matters most is to concentrate the analysis on a ranked list. Peaks should be ranked according to metrics such as scores or q-values adapted for evaluating read enrichment. The generally accepted p-/q-value threshold of 0.05 does not apply well to genomics data for which thousands of regions are being tested and a minimum threshold of

10-5 up to 10-30 is more adapted to ChIP-seq peaks. Fold enrichment being relative to the signal in the input sample is not a good measure to rank peaks (e.g. the same 2-fold enrichment can come from 2/1 or 10/5 where the absolute count in the ChIP sample and therefore peak height is five times higher in the second fraction). However, it can be used to set a minimal threshold, 2-fold being generally accepted but 5-fold being more adapted to ChIP-seq peaks. FDRs defined empirically are also not well adapted for defining ranks since the very best peaks found in the ChIP sample can have a lower FDR than the next ones in case the input contains just one high peak [29]. Again, it can still be used to set a minimal threshold, 5% being a generally accepted threshold but 1% being more adapted to ChIP-seq peaks. More importantly, the difficulty of choosing thresholds can be overcome by comparing ChIP-seq samples to each other, either within replicated samples or across different conditions, which will be discussed in Chapter 8.

6.3 EXISTING TOOLS AND CONSIDERATIONS

The ChIP-seq assay was introduced in 2007 and many peak callers were developed in subsequent years (reviewed in [30]). It includes MACS [31], which has become the most popular peak caller to date and is still actively supported, and SPP [8], which is among the ones used in the ENCODE pipelines [27]. However, those peak finders were developed on the very first ChIP-seq datasets generated and do not always adapt well to current ChIP-seq datasets that take advantage of recent methodological improvements such as paired-end sequencing, high sequencing depth and most importantly, an increased experimental resolution [6, 32].

6.3.1 Single-end versus paired-end libraries

In ChIP-seq experiments, since the fragment size can be estimated from single-end data, the performance in finding peaks is only slightly improved with the use of paired-end over single-end data [6]. Therefore, most ChIP-seq datasets have been generated using single-end libraries and some peak callers do not offer the possibility to use paired-end data. When comparing paired-end with single-end datasets, paired-end can also be treated as a single-end input (using only one of the two sets).

6.3.2 Sequencing depth and library complexity

The first ChIP-seq datasets generated had about 2 to 5 million (M) sequenced reads [33], whereas recent ones have about 20 to 50 M reads, which resulted in a 10-fold increase in sequencing depth over ten years. A good sequencing depth is crucial to be able to identify all true binding sites in a sample and can be assessed by performing a saturation analysis (see Section 6.6). However, since reads mapping to the same position can also arise from PCR amplification artefacts, the total number of mapped reads does not necessarily reflect the library complexity (see Chapter 4). Some peak callers therefore use a step of duplicate removal before calculating read enrichments. While this strategy is valid for datasets with few million reads, where duplicates are mainly PCR artefacts, nowadays, high sequencing depth implies that more reads mapping to the same position could indeed come from true distinct DNA fragments [6]. Therefore, if the ChIP-seq library is of good quality and shows high complexity, we do not recommend the removal of duplicated reads for peak calling. Some peak callers (e.g. MACS2) now have an option to remove only a fraction of the duplicated reads based on an estimation of the genuine duplicates using the sequencing depth and the mappable genome size.

6.3.3 Experimental resolution

Peak callers typically merge close peaks into large regions, which may result in a loss of resolution. Therefore, it is important to separate the two types of regions examined by ChIP-seq: broad versus sharp. On the one hand, data from histone ChIP-seq show broad enrichment of signal and require identification of large regions (example code in Section 6.5). Peak finders for histone marks should therefore focus on the identification of region boundaries rather than the determination of peak summits. It is a non-trivial task that is still a challenge and current peak callers still struggle to find consistent histone-bound region boundaries across samples. This might also be due to the nature of this type of signal.

On the other hand, data from TF ChIP-seq show sharp signal enrichment, and need methods that identify narrow regions (example code in Section 6.4). Since regulatory regions consist of multiple TF binding sites [34, 35], the resolution of the ChIP-seq data is crucial to distinguish them individually. In recent methodological improvements such as ChIP-exo [36] and ChIP-nexus [3] the experimental resolution was increased to a few base pairs. Therefore, peak callers applied to TF ChIP-seq data should be adapted for these new methods that do not have controls and make use of the experimental resolution by focusing on peak summits instead of whole peak regions. To determine which methods work best for TF data, the best indicator is the minimal distance between peak summits and not the size of the peak regions [32].

6.3.4 New generation of peak callers

Methods developed more recently did consider some of the specifics discussed above. The updated version of MACS [31], MACS2, identifies both broad and sharp peaks (example code in Section 6.5). Peakzilla [32] was developed specifically to identify peaks from TF ChIP-seq data at high resolution (example code in Section 6.4). HOMER [37], originally designed as a tool to identify motifs de novo in peak regions (example code in Chapter 9), also developed other tools such as findPeaks to call peaks from different types of ChIP-seq. JAMM [38] uses replicated samples to improve the resolution and precision of peak width. GEM [39] identifies peaks at high resolution by including information about TF motifs as an additional input, however, we prefer to use motif information to validate peaks rather than at the identification step. Additional methods have been developed specifically to compare different samples, which will be discussed in Chapter 8.

6.3.5 Post-processing

Post-processing steps can be applied to remove artefactual peaks that cannot be filtered during the peak calling procedure. Such peaks are in blacklisted genomic regions that show very high enrichment of reads in any ChIP-seq dataset (ChIP and input), often located at centromeres and telomeres, and have been defined by ENCODE for several species (see Chapter 2.2.2). We also chose to remove peaks located on the mitochondrial chromosome (chrM), which often shows biased constant read enrichment.

6.4 PEAKZILLA FOR TRANSCRIPTION FACTOR DATA

Peakzilla [32] was designed specifically for sharp TF ChIP-seq data to call peaks at high resolution i.e. resolve closely spaced peak summits resulting from homotypic binding of TFs. Peakzilla can be used on ChIP-exo data as well, i.e. without a control sample. Briefly, it estimates the fragment size using the double distribution of forward and reverse reads from highly enriched regions. Then, it directly scans the double distribution of reads along the genome using a sliding window to score regions. To do so, it first calculates the number of reads in the ChIP samples subtracted by the number of reads in the control sample (normalised by the total number of mapped reads in the library). This method allows for a better ranking than fold changes. It then weighs this raw score with a p-value that checks how well the data fit the expected double Gaussian distribution of forward and reverse reads. This allows filtering out positions with PCR artefacts without removing duplicate reads, as well as better identifying the exact summit position. Finally, it calculates a fold enrichment for each region and corrects for multiple testing by calculating an empirical FDR through swapping ChIP and control samples. As default, it returns peaks with a minimum score of 1 and fold enrichment of 2. The only inputs needed are the two ChIP and control files.

```
# Create directory for peaks
mkdir peaks

sample=NRF1_CHIP_WT_1
control=NRF1_INPUT_WT

## Break down of the commands (many intermediate files)

# Run Peakzilla
bamToBed -i reads/${sample}.bam > reads/${sample}.bed
bamToBed -i reads/${control}.bam > reads/${control}.bed
peakzilla.py reads/${sample}.bed reads/${control}.bed -l
peaks/${sample}_peakzilla_report.txt >
peaks/${sample}_peakzilla.tsv

# Removal of peaks on chrM (density is usually abnormally
high along the complete chrM)
# Reorganisation of output into BED-like format: chromosome
/ start (0-based) / end / summit / score / fold_enrichment
awk -v OFS="\t" '(NR>1&&$1!="chrM"){print
$1,$2-1,$3,$5,$6,$9}' peaks/${sample}_peakzilla.tsv >
peaks/${sample}_peaks_peakzilla_tmp.bed


# Removal of peaks in blacklisted regions
```

```
intersectBed -v -a peaks/${sample}_peaks_peakzilla_tmp.bed
-b mm10_blacklist.bed.gz > peaks/${sample}_peakzilla.bed

# Remove intermediate files
rm reads/${sample}.bed reads/${control}.bed
peaks/${sample}_peakzilla.tsv
peaks/${sample}_peaks_peakzilla_tmp.bed

## One-line command (~6 min)
peakzilla.py < (bamToBed -i reads/${sample}.bam) <
(bamToBed -i reads/${control}.bam) -1
peaks/${sample}_peakzilla_report.txt | awk -v OFS="\t"
'(NR>1&&$1!="chrM"){print $1,$2-1,$3,$5,$6,$9}' |
intersectBed -v -a stdin -b mm10_blacklist.bed.gz >
peaks/${sample}_peaks_peakzilla.bed
```

6.5 MACS2 FOR HISTONE MARK DATA

MACS [31] was one of the first peak finders for ChIP-seq data
and remains the most popular one. Initially, it was developed to
identify sharp peaks, but the defined peaks were relatively wide.
The latest version, MACS2, has now two different functions for narrow and broad peaks. Briefly, it first removes all duplicated reads.
Then the fragment size is estimated using the double distribution
of forward and reverse reads from highly enriched regions. It ex-
tends all reads to the estimated fragment size and scales ChIP and
input to the same sequencing depth i.e. normalises to the total
number of mapped reads in the library. It then scans the distribution of fragments along the genome using a sliding window to
score regions. To do so, it compares the enrichment in ChIP versus
control samples and calculates a significance score using a local
Poisson distribution. Finally, it corrects for multiple testing using
the Benjamini–Hochberg procedure. As default, it returns peaks
with a minimum q-value score of 0.05. Input parameters are the
path to the ChIP and control files, the file format, the genome size,
an output directory, and a sample name. The option to call broad
peaks is --broad.

```
# Run MACS2 (~7 min)
# q-value threshold can be adapted with the option
--broad-cutoff (default 0.1)
macs2 callpeak -t reads/${sample}.bam -c
reads/${control}.bam -f BAM -g mm --outdir peaks -n
${sample} --broad 2> peaks/${sample}_macs2_report.txt
```

```
# Reorganisation of output into bed-like format: chromosome
/ start (0-based) / end / summit / score / fold_enrichment
cat peaks/${sample}_peaks.xls | grep -v "#" | awk -v
OFS="\t" '(NR>2&&$1!="chrM"){print $1,$2,$3,$4,$6,$7}' |
intersectBed -v -a stdin -b mm10_blacklist.bed.gz >
peaks/${sample}_peaks_macs.bed

# Removal of intermediate files
rm peaks/${sample}_peaks.xls peaks/${sample}_model.r
peaks/${sample}_peaks.broadPeak
peaks/${sample}_peaks.gappedPeak
```

6.6   SATURATION ANALYSIS

In order to check that the sample was sequenced deep enough to identify most bound regions, it is advisable to perform a saturation analysis. It involves randomly subsampling the number of reads and calculating the number of peaks identified using those subsets. The number of peaks are then plotted against the number of reads used. If the number of peaks show saturation and reach a plateau, then the sample was sequenced deep enough. NRF1 ChIP-seq samples show saturation at ~20 million reads (Figure 6.2).

```
sample=NRF1_CHIP_WT_1
control=NRF1_INPUT_WT

## Number of peaks found
wc -l peaks/${sample}_peaks_peakzilla.bed

## Saturation analysis
# Does the number of peaks saturate when using all
available reads
NB=$(samtools idxstats reads/${sample}.bam | awk
'{t=t+$3}END{print t}')
L=$sample
for subset in 'seq 1000000 1000000 $NB';do
    L=$L"\t"$(peakzilla.py < (bamToBed -i
    reads/${sample}.bam | shuf -n $subset
    --random-source=reads/${sample}.bam) < (bamToBed -i
    reads/${control}.bam) -l /dev/null | wc -l)
done
echo -e $L > peaks/${sample}_peakzilla_saturation_table.txt
```

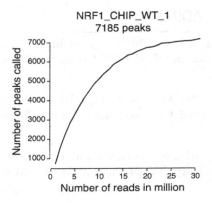

NRF1_CHIP_WT_1
7185 peaks

Figure 6.2 **Saturation analysis of NRF1 ChIP-seq.** The number of peaks called using each random read subset is plotted against the read count in the given subset. The number of peaks using all data is indicated above.

```
# Open R
R

sample="NRF1_CHIP_WT_1"

# Load saturation table
d=read.table(paste("peaks/",sample,
"_peakzilla_saturation_table.txt",sep=""),row.names=1)

# Generate the saturation plot
pdf(paste("peaks/",sample,"_peakzilla_saturation_plot.pdf",
sep=""))
par(bg="white")
plot(seq(1,ncol(d),1),d[1,],type="l",xlab="Number of reads
in million",ylab="Number of peaks
called",main=c("Saturation
analysis",rownames(d)[1],paste("Total",d[1,ncol(d)],
"peaks")))
dev.off()

# Close R
q()
```

FURTHER READING

Landt S.G. et al. (2012). ChIP-seq guidelines and practices of the ENCODE and modENCODE Consortia. *Genome Res*, 22(9): 1813–1831.

Bailey T. et al. (2013). Practical guidelines for the comprehensive analysis of ChIP-seq data. *PLoS Comput Biol*, 9(11): e1003326.

Chen Y. et al. (2012). Systematic evaluation of factors influencing ChIP-seq fidelity. *Nat Methods*, 9(6): 609–614.

## CHAPTER 7

Data Visualisation

Anaïs Bardet

V ISUALISATION of the data generated throughout the analysis is essential to get an impression of data quality, understand the results and integrate the data with other information. Additionally, it allows focusing on interesting candidates. In this chapter, we will generate files that enable the visualisation of the data at read or peak level in genome browsers.

7.1 READ DENSITIES

Visualisation of the reads can already be carried out after genomic alignment (see Chapter 4). BAM files containing the aligned reads can be loaded directly into genome browsers (reads mapping to the forward or reverse strand appear distinctively) (Figure 7.1). However, in regions with high read density such as peaks, where hundreds of reads align, such a visualisation does not scale well. This approach is therefore useful only to investigate information at the level of individual reads (e.g. exact positions, mismatches, etc).

Instead of visualising each single read, a genomic coverage file can be generated to summarise the number of reads overlapping each position in the genome (Figure 7.1). Since the sequenced reads only represent the ends (e.g. 50 bp) of the co-purified DNA fragments bound by the TF of interest, each read is extended to 200 bp in the 3' direction. 200 bp represents a rough estimate of the fragment size selected experimentally in the ChIP-seq protocol (see Chapter 1.1). This smooths the signal and ensures that the summits of the peaks appear above the binding sites. To be able to compare different samples, read counts at each position are normalised

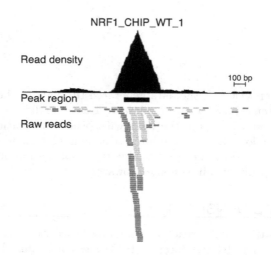

Figure 7.1 **Data visualisation in a genome browser.** Visualisation of a peak with score 3 from the sample NRF1_CHIP_WT_1 centred at position chr2:153652399 on the mm10 genome. Read density, peak region and raw reads mapping to the forward (dark grey) or reverse strand (light grey) are shown as generated by the UCSC genome browser.

to one million of mapped reads using a scale factor (1,000,000 / number_mapped_reads). Genomic coverage of reads is stored in the bedGraph format (chromosome / start / end / read_density) and can be compressed into a bigWig format.

```
# Create a directory for the density tracks
mkdir tracks

sample=NRF1_CHIP_WT_1

# Chromosome sizes (to calculate read densities at all
positions)
# To be downloaded from UCSC (genome.ucsc.edu/), Downloads,
Genome Data
# File is sorted according to chromosome and only
conventional chromosomes are kept
wget
http://hgdownload.cse.ucsc.edu/goldenPath/mm10/bigZips/mm10.
chrom.sizes sort -k1,1 mm10.chrom.sizes | grep -v chrUn |
grep -v random > mm10.chrom.sizes.tmp
mv mm10.chrom.sizes.tmp mm10.chrom.sizes

# Scale to normalise the number of reads at each position
to 1,000,000 mapped reads in the library
# Count total number of mapped reads if the index BAI file
is available (~1 min)
samtools idxstats reads/${sample}.bam | awk
'{if($1!="*"){total=total+$3}}END{print 1000000/total}'
# Otherwise count total number of mapped reads from the BAM
file (~ 2 min)
bamToBed -i reads/${sample}.bam | wc -l | awk '{print
1000000/total}'
scale=0.0322351
# One-line command
scale=$(samtools idxstats reads/${sample}.bam | awk
'{if($1!="*"){total=total+$3}}END{print 1000000/total}')

# Extend reads checking they do not extend over chromosome
limits
extend=200
# Extend reads from the forward strand (since files are
sorted by starts, there is no need to sort again)
bamToBed -i reads/${sample}.bam | awk -v OFS="\t" -v
E=$extend -v file=mm10.chrom.sizes
'BEGIN{while(getline<file){S[$1]=$2}}($6=="+"){if($2+E>S[$1])
```

```
{print $1,$2,S[$1]}else{print $1,$2,$2+E}}' >
tracks/${sample}_forward.bed
# Extend reads from the reverse strand (since ends are not
sorted if reads do not have the same length e.g. after
trimming, we need to sort the output again by chromosome /
start)
bamToBed -i reads/${sample}.bam | awk -v OFS="\t" -v
E=$extend '($6=="-"){if($3-E<1){print $1,"1",$3}else{print
$1,$3-E,$3}}' | sort -k1,1 -k2,2n >
tracks/${sample}_reverse.bed
# Merge sorted file (sort -m much faster than just sort
since input files are already sorted)
sort -m -k1,1 -k2,2n tracks/${sample}_forward.bed
tracks/${sample}_reverse.bed > tracks/${sample}.bed

# Generate bedGraph file from BED (read coverage at each
non-zero position)
genomeCoverageBed -bg -i tracks/${sample}.bed -g
mm10.chrom.sizes -scale $scale > tracks/${sample}.bg
# Remove intermediate files
rm tracks/${sample}_forward.bed
tracks/${sample}_reverse.bed tracks/${sample}.bed

# One-line command (~6 min)
extend=200
sort -m -k1,1 -k2,2n <(bamToBed -i reads/${sample}.bam |
awk -v OFS="\t" -v E=$extend -v file= mm10.chrom.sizes
'BEGIN{while(getline<file){S[$1]=$2}}($6=="+"){if($2+E>S[$1])
{print $1,$2,S[$1]}else{print $1,$2,$2+E}}') <(bamToBed -i
reads/${sample}.bam | awk -v OFS="\t" -v E=$extend
'($6=="-"){if($3-E<1){print $1,"1",$3}else{print
$1,$3-E,$3}}' | sort -k1,1 -k2,2n) | genomeCoverageBed -bg
-i stdin -g mm10.chrom.sizes -scale $scale >
tracks/${sample}.bg

# Generate bigWig file from bedGraph (~2 min)
bedGraphToBigWig tracks/${sample}.bg mm10.chrom.sizes
tracks/${sample}.bw

# Remove intermediate file
rm tracks/${sample}.bg
```

7.2 PEAK REGIONS

Visualisation of the peak regions can be performed after the peak calling step (see Chapter 6). Simple BED files (chromosome / start / end) corresponding to the peak regions can be used directly or can also be compressed in a bigBed format.

```
# Generate a simple BED file (chromosome / start / end)
cut -f1-3 peaks/${sample}_peaks_peakzilla.bed >
peaks/${sample}_peaks_peakzilla_short.bed
```

```
# Generate bigBed file from BED
bedToBigBed peaks/${sample}_peaks_peakzilla_short.bed
mm10.chrom.sizes tracks/${sample}_peaks_peakzilla.bb
rm peaks/${sample}_peaks_peakzilla_short.bed
```

Note: Exploring the best and worst ranked peaks in a genome browser allows us to assess the selected threshold, and check if the peaks are shared in other samples. However, it is important to keep in mind that this visualisation does not consider all underlying characteristics of the data. Therefore, it is not worth adjusting initial peak calling parameters to fit any one given region as it might not work well for others.

7.3 GENOME BROWSER

The University of California Santa Cruz (UCSC) genome browser [40] is available online or can be installed locally. The Integrative Genomic Viewer (IGV) [41] developed at the Broad Institute is a popular alternative that can be run locally. This is particularly useful when working without access to a dedicated server with a web server.

Once the reference genome has been selected in UCSC, various pre-loaded data tracks are available (e.g. gene annotations). Additional datasets can be uploaded manually using the 'My Data' section under 'Custom Tracks'. Standard data formats such as bedGraph or BED can be used. Since the complete files are uploaded to the genome browser, this step can be slow for large files. As a faster alternative , it is recommended to use compressed files in indexed binary formats such as bigWig or bigBed, from which only the data corresponding to the region of interest will be loaded into the browser. Therefore those files need to be accessible from a web server (http, https, or ftp) with a URL and can be loaded in UCSC by encapsulating the URL in the following manner:

```
# BAM files (with .bai file in same directory)
# track type=bam name="NRF1_CHIP_WT_1_reads"
bigDataUrl=http://your_server/NRF1_CHIP_WT_1.bam

# bigWig files
# track type=bigWig name="NRF1_CHIP_WT_1_density"
bigDataUrl=http://your_server/NRF1_CHIP_WT_1.bw

# bigBed files
# track type=bigBed name="NRF1_CHIP_WT_1_peaks" bigDataUrl=
http://your_server/NRF1_CHIP_WT_1_peaks_peakzilla.bb
```

> Note: The data associated with this book, including URLs to be loaded in the UCSC genome browser, are available online (anaisbardet.cnrs.fr/ practical-guide-to-chip-seq-data-analysis/).

Tracks loaded in the genome browser are then listed at the bottom of the page and can be shown in different formats (full, pack, squish or dense) or be hidden. They can also be reorganised by dragging the grey vertical bar on the left side of each track or configured by right-clicking this grey bar. For example, a common scale for all samples starting at zero can be set for density tracks.

To avoid having to load the same tracks and their configuration each time we want to look at the data, track hubs can be generated and loaded using the 'My Data' section under 'Track Hubs'. Information about track hubs can be found in the UCSC help (genome-euro.ucsc.edu/goldenPath/help/ hgTrackHubHelp.html).

## CHAPTER 8

Comparative Analysis

Anaïs Bardet and Borbala Mifsud

C OMPARATIVE ANALYSIS of ChIP-seq data is essential to compare the binding of a protein of interest either between replicate experiments or under different physiological conditions. This chapter will introduce how to quickly compare ChIP-seq samples by overlapping peaks and how to use a more quantitative approach based on read densities within peaks to investigate differential binding.

8.1   OVERLAP OF PEAK REGIONS

A straightforward approach to compare peaks across samples is a simple overlap. This results in a binary view of whether a peak is found in both samples, providing a rough estimate of the similarity in peak regions between different samples. However, this approach intrinsically underestimates the similarity as peak calling relies on a confidence threshold applied on a ranked list of regions (see Chapter 6). Therefore, a peak that passed the threshold in one sample might be just below in a second sample, even though it shows decent enrichment [42]. An alternative would be to overlap high-confidence peaks from the first sample with all peaks identified with a lower confidence threshold in the second sample [42, 43]. Although to a lesser extent, a simple overlap of peaks also underestimates differences between samples, because even if a peak is above the threshold in both samples, it can still show very different enrichment.

The following code shows how to calculate all pairwise overlaps.

```
for sample1 in NRF1_CHIP_WT_1 NRF1_CHIP_WT_2
NRF1_CHIP_TKO_1 NRF1_CHIP_TKO_2
do
    L=$sample1"\t"$(cat peaks/${sample1}_peaks_peakzilla.bed
    | wc -l)
    for sample2 in NRF1_CHIP_WT_1 NRF1_CHIP_WT_2
    NRF1_CHIP_TKO_1 NRF1_CHIP_TKO_2
    do
        L=$L"\t"$(intersectBed -u -a
        peaks/${sample1}_peaks_peakzilla.bed -b
        peaks/${sample2}_peaks_peakzilla.bed | wc -l)
    done
    echo -e $L
done > changes/peak_overlap_table.txt

# Table with: sample / total_number_of_peaks /
overlap_for_each_sample
cat changes/peak_overlap_table.txt | column -t

# Transformation into percentages
cat changes/peak_overlap_table.txt | awk
'{L=$1"\t"$2;for(i=3;i<=NF;i++){L=L"\t"$i*100/$2};print L}'
> changes/peak_overlap_percent_table.txt
cat changes/peak_overlap_percent_table.txt | column -t

# Clustering of samples and representation in a heatmap

# Open R and load library
R
library(NMF)

# Load table
x=read.table("changes/peak_overlap_percent_table.txt")
data=x[,3:6]
rownames(data)=colnames(data)=x[,1]

# Generate a heatmap that clusters the samples
pdf("changes/peak_overlap_percent_heatmap.pdf")
aheatmap(data,breaks=50,annRow=data.frame(condition=c("WT",
"WT","TKO","TKO"),replicate=as.factor(c(1,2,1,2))),
annColors=list('condition'=c("blue","red"),
'replicate'=c("black","grey")))
dev.off()
```

```
# Close R
q()
```

This binary approach shows some limitations. Replicate samples differ in the number of peaks identified with the default `peakzilla` threshold (7167 versus 10232). Therefore, the analysis is not symmetrical: 98% of the peaks in the WT replicate 1 overlap with replicate 2, but only 67% of the peaks in replicate 2 overlap with replicate 1. However, the additional peaks in replicate 2 might have very well been present in replicate 1 if considering additional enriched regions. When the sample is not saturated, deeper sequencing could also increase replicate overlap. Differences in the length of peak regions can further distort the result, especially if any overlap, down to just 1 bp, is accepted.

Note: Because of these biases, peak overlaps commonly shown in Venn diagrams can be misleading, as non-overlapping peaks should not be interpreted as sample-specific.

Despite these limitations, we can conclude in the present example that the two replicate experiments for NRF1 in WT and TKO cells seem to share most of their peak regions. As expected, we observe fewer peaks that overlap between WT and TKO conditions. Nevertheless, there is still a substantial overlap (76% of WT or 45% of TKO), suggesting that many peaks are shared between conditions.

The resulting heatmap of all pairwise comparisons allows us to examine which samples are more similar than others (Figure 8.1A). As expected, the samples cluster by condition. Additionally, WT peaks seem to be more often shared with TKO peaks than vice versa, although this might be an artefact due to the higher peak number in the TKO samples.

8.2   IRREPRODUCIBLE DISCOVERY RATE (IDR)

As mentioned above, every comparison in ChIP-seq analysis strongly depends on the thresholds that were applied during the peak calling step. As a consequence, the ENCODE project developed the irreproducible discovery rate (IDR), as a metric to identify genuine peaks based on their reproducibility across replicates [44]. The underlying idea is that a peak list generated with a lenient threshold will contain both genuine peaks and noise. When

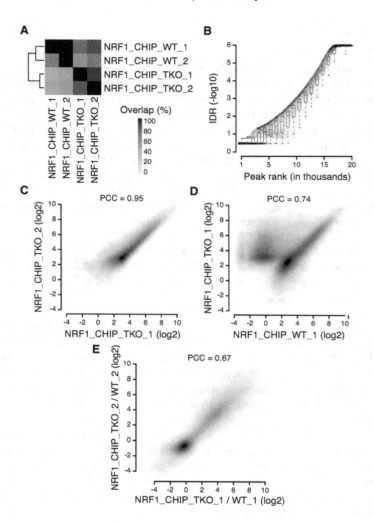

Figure 8.1 **Comparison of peak regions.** (**A**) Heatmap of the overlap of WT and TKO peak regions. (**B**) Peak rank heterogeneity ($-\log_{10}$IDR) is shown as a function of peak rank. Boxplots show the summary for bins of 1,000 peaks each. Scatterplots comparing read densities in peak regions between TKO replicate samples (**C**) or between WT and TKO samples (**D**). (**E**) Scatterplot comparing ratios of read densities in peak regions between TKO and WT for both replicates.

the peaks in the list are ranked, e.g. on their fold enrichment or significance, these ranks will correlate well for genuine peaks, while the noise will show no correlation. IDR uses a statistical method to find the point in the curve at which the heterogeneity of the association between replicate ranks sharply increases (Figure 8.1B).

Note: This method can also be used on peak lists from the same sample generated by different peak callers. In absence of replicates, this approach can help to define reliable peaks within a single sample.

Note: IDR is not applicable to histone marks, because of the ambiguous overlap of broad peaks.

8.2.1 Peak calling for IDR

IDR can be used with the output of any peak caller that ranks the peaks, as long as the scoring does not produce too many ties which would lead to ambiguous ranking. Here we show how to use IDR with `peakzilla` peaks. As mentioned before, for IDR to work, the list of peaks has to contain both genuine peaks and noise, and therefore the peak calling parameters have to be relaxed.

When comparing replicates, a unified set of peaks is needed as a reference set. This can either be created by the IDR software from the separately identified peak regions or supplied by the user. For the latter, pooling the replicates before peak calling should help in identifying all possible peak regions, and the peak list from the merged sample is commonly used for IDR analysis.

```
# Call peaks on each replicate with a relaxed threshold (-s
0.1 instead of default 1)
peakzilla.py -s 0.1 <(bamToBed -i reads/NRF1_CHIP_WT_1.bam)
<(bamToBed -i reads/NRF1_INPUT_WT.bam) -l
peaks/NRF1_CHIP_WT_1_all_peakzilla_report.txt | awk -v
OFS="\t" '(NR>1&&$1!="chrM"){print $1,$2-1,$3,$5,$6,"+"}' |
intersectBed -v -a stdin -b mm10_blacklist.bed.gz >
peaks/NRF1_CHIP_WT_1_all_peaks_peakzilla.bed
peakzilla.py -s 0.1 <(bamToBed -i reads/NRF1_CHIP_WT_2.bam)
<(bamToBed -i reads/NRF1_INPUT_WT.bam) -l
peaks/NRF1_CHIP_WT_2_all_peakzilla_report.txt | awk -v
OFS="\t" '(NR>1&&$1!="chrM"){print $1,$2-1,$3,$5,$6,"+"}' |
intersectBed -v -a stdin -b mm10_blacklist.bed.gz >
peaks/NRF1_CHIP_WT_2_all_peaks_peakzilla.bed
```

```
# Pool BAM files and call peaks on merged file with a
relaxed threshold.
samtools merge reads/NRF1_CHIP_WT.bam
reads/NRF1_CHIP_WT_1.bam reads/NRF1_CHIP_WT_2.bam

peakzilla.py -s 0.1 <(bamToBed -i reads/NRF1_CHIP_WT.bam)
<(bamToBed -i reads/NRF1_INPUT_WT.bam) -l
peaks/NRF1_CHIP_WT_1_all_peakzilla_report.txt | awk -v
OFS="\t" '(NR>1&&$1!="chrM"){print $1,$2-1,$3,$5,$6,"+"}' |
intersectBed -v -a stdin -b mm10_blacklist.bed.gz >
peaks/NRF1_CHIP_WT_all_peaks_peakzilla.bed
```

8.2.2 Calculating IDR

The next step is to run the IDR software to identify the reproducible peaks. For biological replicates, a commonly applied threshold is 0.05, meaning that up to 5% of final list will not be reproducible. A lower threshold should be used for technical replicates because of their lower overall variability.

```
idr --samples peaks/NRF1_CHIP_WT_1_all_peaks_peakzilla.bed
peaks/NRF1_CHIP_WT_2_all_peaks_peakzilla.bed --peak-list
peaks/NRF1_CHIP_WT_all_peaks_peakzilla.bed
--input-file-type bed --rank score --idr-threshold 0.05
--output-file peaks/NRF1_CHIP_WT_idr --plot
```

The concept of IDR relies heavily on having two good replicates. If one of the replicates shows poorer quality, e.g. if the immunoprecipitation was not efficient, IDR will only record very few reproducible peaks. For this case, ENCODE developed a rescue strategy by pooling both replicates and then randomly splitting the reads into two pseudo-replicates. These do not represent true biological or experimental variation, but serve to model the stochastic noise in the sampling of reads from a population of DNA fragments. For IDR comparison of pseudo-replicates, it is suggested to use a lower threshold (0.0025) due to the reduced noise.

8.3 COMPARISON OF READ DENSITIES

At the level of read densities, the most simplistic approach to assess the overall similarity between samples is to globally correlate their read densities. This can be achieved by calculating a Pearson correlation coefficient (PCC) on normalised read densities across the genome, either for every individual base pair [42] or in a sliding window along the genome. However, since the peak regions represent only a minor fraction of the genome, a high correlation of read densities will mostly reflect a consistent background signal, as indicated by the high correlation between ChIP and input samples (see Table 8.1).

TABLE 8.1 Pearson correlation coefficients (PCC) of read densities.

Sample 1	Sample 2	PCC
NRF1_CHIP_WT_1	NRF1_CHIP_WT_2	0.98
NRF1_CHIP_WT_1	NRF1_CHIP_TKO_1	0.97
NRF1_CHIP_WT_1	NRF1_INPUT_WT	0.96

PCC of read densities for every base pair along the genome (excluding positions with zero reads in both samples).

To compare the read densities specifically in peak regions, a PCC can be calculated only within regions that contain a peak in at least one of the samples. Normalised mean read densities for each region can also be compared visually using a scatterplot. This represents a more quantitative comparison of the signal in peak regions as opposed to the binary approach of overlapping peak regions explained in Section 8.1.

8.3.1 Merging peak regions

To focus on genomic windows that contain high signal and might differ across samples, we perform the comparison of read densities within peak regions. To do so, we merge peak regions from all experiments (could be performed for all possible pairwise comparisons separately). Even though this also relies on peak calling thresholds, it allows quantitative comparison of all peaks, including those that are present in only one sample.

```
# Concatenate peaks from all samples and sort by chr and
start
cat peaks/NRF1_CHIP_WT_1_peaks_peakzilla.bed
peaks/NRF1_CHIP_WT_2_peaks_peakzilla.bed
peaks/NRF1_CHIP_TKO_1_peaks_peakzilla.bed
peaks/NRF1_CHIP_TKO_2_peaks_peakzilla.bed | sort -k1,1
-k2,2n > changes/NRF1_all_regions_tmp.txt

# Merge peak regions from all samples
mergeBed -i changes/NRF1_all_regions_tmp.txt >
changes/NRF1_all_regions.txt

# Remove intermediate files
```

```
rm changes/NRF1_all_regions_tmp.txt

# One-line command
cat peaks/NRF1_CHIP_WT_1_peaks_peakzilla.bed
peaks/NRF1_CHIP_WT_2_peaks_peakzilla.bed
peaks/NRF1_CHIP_TKO_1_peaks_peakzilla.bed
peaks/NRF1_CHIP_TKO_2_peaks_peakzilla.bed | sort -k1,1
-k2,2n | mergeBed -i stdin > changes/NRF1_all_regions.txt
```

8.3.2 Counting reads for each sample

We overlap the merged regions with the original reads to count the number of reads that fall into them in each sample. These raw counts will be used directly as input to identify differential peak regions in the next section.

```
# For one sample
intersectBed -c -sorted -a changes/NRF1_all_regions.txt -b
reads/NRF1_CHIP_WT_1.bam >
changes/NRF1_all_regions_count1.txt

# For all samples (~2 min)
intersectBed -c -sorted -a changes/NRF1_all_regions.txt -b
reads/NRF1_CHIP_WT_1.bam | intersectBed -c -sorted -a stdin
-b reads/NRF1_CHIP_WT_2.bam | intersectBed -c -sorted -a
stdin -b reads/NRF1_CHIP_TKO_1.bam | intersectBed -c
-sorted -a stdin -b reads/NRF1_CHIP_TKO_2.bam >
changes/NRF1_all_regions_count.txt
cat changes/NRF1_all_regions_count.txt | head | column -t
```

8.3.3 Normalising read counts

Since peak regions have different sizes and samples contain different numbers of mapped reads, each read count is normalised to the size of the region and the total number of mapped reads in the sample to obtain RPKM values (reads per kilobase per million).

```
# Normalised read counts (RPKM table)
# RPKM = reads x (1000000 / total_library) x (1000 /
region_size)
TOT1=$(samtools idxstats reads/NRF1_CHIP_WT_1.bam | awk
'($1!~"*"){t=t+$3}END{print t}')
TOT2=$(samtools idxstats reads/NRF1_CHIP_WT_2.bam | awk
'($1!~"*"){t=t+$3}END{print t}')
```

```
TOT3=$(samtools idxstats reads/NRF1_CHIP_TKO_1.bam | awk
'($1!~"*"){t=t+$3}END{print t}')
TOT4=$(samtools idxstats reads/NRF1_CHIP_TKO_2.bam | awk
'($1!~"*"){t=t+$3}END{print t}')
cat changes/NRF1_all_regions_count.txt | awk -v OFS="\t"
-vTOT1=$TOT1 -vTOT2=$TOT2 -vTOT3=$TOT3 -vTOT4=$TOT4
'{size=$3-$2;print
$1,$2,$3,$4*(1000000/TOT1)*(1000/size),$5*(1000000/TOT2)*
(1000/size),$6*(1000000/TOT3)*(1000/size),$7*(1000000/TOT4)*
(1000/size)}' > changes/NRF1_all_regions_rpkm.txt
cat changes/NRF1_all_regions_rpkm.txt | head | column -t

# Count reads and generate RPKM tables using for loop (~6
min)
cat changes/NRF1_all_regions.txt >
changes/NRF1_all_regions_count.txt
cat changes/NRF1_all_regions.txt >
changes/NRF1_all_regions_rpkm.txt
for sample in NRF1_CHIP_WT_1 NRF1_CHIP_WT_2 NRF1_CHIP_TKO_1
NRF1_CHIP_TKO_2
do
    TOT=$(samtools idxstats reads/${sample}.bam | awk
    '($1!~"*"){t=t+$3}END{print t}')
    intersectBed -c -sorted -a
    changes/NRF1_all_regions_count.txt -b
    reads/${sample}.bam > tmp
    mv tmp changes/NRF1_all_regions_count.txt
    paste changes/NRF1_all_regions_rpkm.txt <(cat
    changes/NRF1_all_regions_count.txt | awk -v OFS="\t"
    -vTOT=$TOT '{size=$3-$2;print
    $NF*(1000000/TOT)*(1000/size)}') > tmp
    mv tmp changes/NRF1_all_regions_rpkm.txt
done
cat changes/NRF1_all_regions_count.txt | head | column -t
cat changes/NRF1_all_regions_rpkm.txt | head | column -t
```

8.3.4 Comparing read counts

We can now visualise the normalised read counts in peak regions
across samples in a scatterplot and calculate the associated PCC.
Scatterplots offer an unbiased way to explore the data and to draw
conclusions. To better visualise the spread of the data, RPKM
values are represented on a $\log_2$ scale. To confirm that the changes
in NRF1 binding between WT and TKO cells are reproducible, we

compare the TKO versus WT fold changes between replicates. To this end, we add a pseudo-count (here 0.1) to all data points to avoid division by 0 and represent the values on a $\log_2$ scale to get a normal distribution centred around 0.

```
# Open R
R

# Load RPKM table
data=read.table("changes/NRF1_all_regions_rpkm.txt")
colnames(data)=c("chr","start","end","NRF1_CHIP_WT_1",
"NRF1_CHIP_WT_2","NRF1_CHIP_TKO_1","NRF1_CHIP_TKO_2")

# Visualise pairwise comparisons in a scatterplot
pdf("changes/NRF1_all_regions_rpkm_cor_scatter.pdf")
par(bg="white",mfrow=c(2,2))
smoothScatter(log2(data$NRF1_CHIP_WT_1),
log2(data$NRF1_CHIP_WT_2),xlim=c(-4,10),ylim=c(-4,10),
xlab="NRF1_CHIP_WT_1 (log2)",ylab="NRF1_CHIP_WT_2
(log2)",main=paste("PCC
=",signif(cor(data$NRF1_CHIP_WT_1,data$NRF1_CHIP_WT_2),2)))
smoothScatter(log2(data$NRF1_CHIP_TKO_1),
log2(data$NRF1_CHIP_TKO_2),xlim=c(-4,10),ylim=c(-4,10),
xlab="NRF1_CHIP_TKO_1 (log2)",ylab="NRF1_CHIP_TKO_2
(log2)",main=paste("PCC
=",signif(cor(data$NRF1_CHIP_TKO_1,data$NRF1_CHIP_TKO_2),2)))
smoothScatter(log2(data$NRF1_CHIP_WT_1),
log2(data$NRF1_CHIP_TKO_1),xlim=c(-4,10),ylim=c(-4,10),
xlab="NRF1_CHIP_WT_1 (log2)",ylab="NRF1_CHIP_TKO_1
(log2)",main=paste("PCC
=",signif(cor(data$NRF1_CHIP_WT_1,data$NRF1_CHIP_TKO_1),2)))
smoothScatter(log2(data$NRF1_CHIP_WT_2),
log2(data$NRF1_CHIP_TKO_2),xlim=c(-4,10),ylim=c(-4,10),
xlab="NRF1_CHIP_WT_2 (log2)",ylab="NRF1_CHIP_TKO_2
(log2)",main=paste("PCC
=",signif(cor(data$NRF1_CHIP_WT_2,data$NRF1_CHIP_TKO_2),2)))
dev.off()

# Compare the RPKM changes between WT and TKO samples
across replicated samples
pdf("changes/NRF1_all_regions_rpkm_cor_delta_scatter.pdf")
par(bg="white")
smoothScatter(log2((data$NRF1_CHIP_TKO_1+0.1)/
(data$NRF1_CHIP_WT_1+0.1)),log2((data$NRF1_CHIP_TKO_2+0.1)/
(data$NRF1_CHIP_WT_2+0.1)),xlim=c(-6,10),ylim=c(-6,10),
```

```
xlab="NRF1_CHIP_TKO_1 / NRF1_CHIP_WT_1
(log2)",ylab="NRF1_CHIP_TKO_2 / NRF1_CHIP_WT_2
(log2)",main=paste("PCC
=",signif(cor((data$NRF1_CHIP_TKO_1+0.1)/
(data$NRF1_CHIP_WT_1+0.1),(data$NRF1_CHIP_TKO_2+0.1)/
(data$NRF1_CHIP_WT_2+0.1)),2)))
dev.off()

# Close R
q()
```

We observe that read densities from the replicate samples for both WT and TKO arrange along the diagonal and correlate very well (PCC >0.9) (Figure 8.1C). When comparing WT with TKO samples, the read densities still correlate but less than between replicates (PCC = 0.74 or 0.81) (Figure 8.1D).

Additionally, we find that all peaks that were identified in the WT samples show similar read densities in both conditions, as all regions with high read densities in the WT sample also have high read densities in the TKO samples and align along the diagonal (Figure 8.1D). In contrast, many peaks that were identified in the TKO samples have low read counts or are not present in the WT samples, visualised by a population of data points in the upper left part of the plot.

Note: There are few or no points with low read densities on the lower left part of the plots, which is due to the fact that we only selected regions that were called peaks in at least one of the samples and therefore show minimal enrichment of read counts. In Figure 8.1C for TKO replicate samples, points in the lower left corner indicate peaks that were identified in the WT samples but have background read densities in the TKO samples. They would not be present if peak regions were merged from a pairwise comparison.

Since both replicates show peaks that are gained in the TKO samples, it is interesting to check whether these are the same in replicate 1 and 2. This is confirmed by comparing the fold changes in a delta-delta plot, which shows that the observed read density changes between WT and TKO are highly consistent between the two replicates (PCC = 0.67) (Figure 8.1E).

8.4   DIFFERENTIAL BINDING ANALYSIS

Once scatterplots confirmed the presence of differential peaks across samples, statistical approaches can define groups of shared or differential peaks for further analysis. There are two major types of tools for differential binding analysis [45]. The first type employs a quantitative approach based on read count data to compare the binding strength in one condition against the other. The second type uses hidden Markov models to segment the genome into lost, unchanged or gained regions. However, these tools do not allow a quantitative description beyond these three distinct states. Here, we introduce a typical analysis workflow for a quantitative approach using DESeq2 as well as DiffBind, which provides a wrapper around DESeq2 specifically for ChIP-seq analysis.

In the following sections, peak regions that show differential binding between conditions are referred to as "WT-specific" or "TKO-specific" if they show more NRF1 binding in WT or TKO cells, respectively. Conversely, peaks that show a less than 2-fold change in binding (in either direction) between conditions are referred to as "shared" peaks.

8.4.1   Using DESeq2

The identification of peak regions with differential enrichment is conceptually similar to the identification of differentially expressed genes, as both rely on the comparison of read counts. This allows us to employ well established statistical methods that were originally designed for RNA-seq data analysis, such as the R/Bioconductor packages DESeq2 [46] and EdgeR [47]. DESeq2 uses negative binomial-based generalised linear models to test the null hypothesis that the $\log_2$ fold change of read counts between two conditions equals to zero. It can be decomposed into four major steps (combined in the function DESeq()):

- Estimation of size factors i.e. normalisation of the raw read counts to the total number of reads in the peak regions (as a proxy of sequencing depth). The size factor is calculated as the median of geometric means of read counts for each peak region.

- Estimation of dispersion i.e. variability between replicates. Since a small number of replicates (often 2-3) results in high variance, DESeq2 uses regions with similar counts to better estimate the dispersion (using an empirical Bayes shrinkage approach).

- Estimation of fold changes. Since ratios of small numbers are inherently noisy, **DESeq2** reduces the fold change estimates for low read counts (using an empirical Bayes shrinkage approach). These so-called moderated fold changes therefore differ from a direct calculation of ratios from the normalised count data.

- Testing for differential enrichment under the null hypothesis that the $\log_2$ fold change of read counts between two conditions is zero. P-values are calculated by a Wald test and adjusted by a Benjamini-Hochberg procedure. An additional filtering uses the mean of normalised counts for all regions to remove adjusted p-values below a given FDR cutoff (default 0.1; introducing NA).

```
# Open R and load library
R
library(DESeq2)

# Load count table (from Section 8.3.2)
x=read.table("changes/NRF1_all_regions_count.txt")
colnames(x)=c("chr","start","end","NRF1_CHIP_WT_1",
"NRF1_CHIP_WT_2","NRF1_CHIP_TKO_1","NRF1_CHIP_TKO_2")
d=x[,4:7]

# Conditions
colData=data.frame(condition=c("WT","WT","TKO","TKO"))
rownames(colData)=colnames(d)

# DESeq2 matrix
dds=DESeqDataSetFromMatrix(countData=d,colData=colData,
design=~condition)
# Filter out regions with no counts (or low counts)
dds=dds[rowSums(counts(dds))>1,]

# Transform data for variance analysis
# Using regularised-logarithm transformation (RLD) when
number of samples < 30
rld=rlog(dds,blind=F)
# Using the variance-stabilising transformation (VST)
otherwise
vsd=vst(dds,blind=F)
```

```
# Principal components analysis (PCA)
pdf("changes/NRF1_all_regions_count_pca.pdf")
plotPCA(rld,intgroup=c("condition"))
dev.off()

# Differential expression (on non-transformed data)
dds=DESeq(dds)

# TKO versus WT
res=results(dds,contrast=c("condition","TKO","WT"))

# Volcano plot (log2 fold change vs. adjusted p-value)
pdf("changes/NRF1_all_regions_count_TKO_vs_WT_volcano.pdf")
par(bg="white")
plot(res$log2FoldChange,-log10(res$padj),xlim=c(-10,10),
ylim=c(0,83),xlab="TKO / WT (log2)",ylab="P-value
(-log10)",pch=20)
lines(c(-10,10),c(10,10))
dev.off()

# Table chr / start / end / NRF1_CHIP_WT_1 / NRF1_CHIP_WT_2
/ NRF1_CHIP_TKO_1 / NRF1_CHIP_TKO_2 / log2FoldChange / padj
y=cbind(x[,1:3],counts(dds,normalized=T),
as.matrix(res)[,c(2,6)])
write.table(y,
"changes/NRF1_all_regions_count_TKO_vs_WT_table.txt",quote=F,
sep="\t",row.names=F,col.names=T)

# Close R
q()

# Select differential genes
# Since the adjusted p-value scales with the fold
enrichment, we only need a p-value threshold
# e.g. 10^-10
cat changes/NRF1_all_regions_count_TKO_vs_WT_table.txt |
head | column -t

# TKO-specific peaks: 2212
# No header, p-value (adjusted) <= 1e-10, log2 fold change
>0 (TKO>WT)
cat changes/NRF1_all_regions_count_TKO_vs_WT_table.txt |
awk '(NR>1&&$9<=1e-10&&$8>0)' >
changes/NRF1_all_regions_count_TKO_spec_table.txt
```

```
# WT-specific peaks: 13
# No header, adjusted p-value <= 1e-10, log2 fold change <0
(WT>TKO)
cat changes/NRF1_all_regions_count_TKO_vs_WT_table.txt |
awk '(NR>1 && $9<=1e-10 && $8<0)' >
changes/NRF1_all_regions_count_WT_spec_table.txt

# Shared peaks: 7068
# No header, absolute log2 fold change < 1 (less than 2
fold change in either direction)
cat changes/NRF1_all_regions_count_TKO_vs_WT_table.txt |
awk '(NR>1 && $8<1 && $8>-1)' >
changes/NRF1_all_regions_count_shared_table.txt
```

A common diagnostic plot in DESeq2 is the principle component analysis (PCA), which allows us to examine the overall similarity between multiple samples (by default using the 500 peak regions with highest variance across samples). In case of the NRF1 ChIP-seq data, this confirms that the replicates cluster together and that 99% of the variance is explained by the changes between the WT and TKO conditions (Figure 8.2A). The volcano plot clearly shows that most differential peaks have increased NRF1 binding in the TKO condition (Figure 8.2B). For the downstream analyses described in Chapter 9, we output a list of TKO-specific and shared peaks (WT-specific peaks are omitted due to their low number). For this, differential peaks are selected according to a stringent significance threshold (adjusted p-value $\leq 10^{-10}$), whereas shared peaks are required to show a less than two-fold change. These selection criteria leave out a grey zone of peaks between those thresholds that allows a clear separation of both groups.

8.4.2 DiffBind

DiffBind is a wrapper tool that applies the R/Bioconductor packages DESeq, DESeq2 or EdgeR to ChIP-seq data (default: DESeq2). It provides a simple workflow and visualises the data at several steps, which allows for checking replicate consistency and overall differences between conditions. It requires a sample sheet, similar to ChIPQC (see NRF1_sample_sheet.csv in Additional Online Data Files), that summarises the information needed about the samples, in a data.frame or csv format.

```
# Open R and load library
```

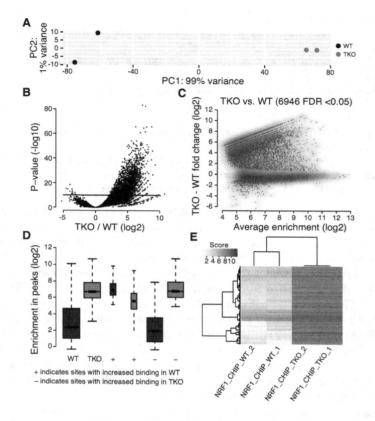

Figure 8.2 **Differential binding analysis.** (**A**) Principal component analysis of WT and TKO ChIP-seq samples. (**B**) Volcano plot showing the -$\log_{10}$ p-value versus the $\log_2$ fold enrichment of the merged peaks. Horizontal line at 10 marks the significance threshold used. (**C**) MA-plot of TKO versus WT. For each peak, the $\log_2$-transformed fold change in TKO versus WT is shown, in relation to the average $\log_2$ fold enrichment across samples. Peaks with significant differential NRF1 binding (FDR<0.05) are overlaid over the distribution of all peaks represented as a density. (**D**) Boxplot of enrichment in all peaks, peaks with decreased and increased NRF1 binding in TKO cells. (**E**) Clustered heatmap showing the enrichment in WT and TKO cells for differentially bound peaks.

```
R
library(DiffBind)

# Load data (~2 min)
NRF1=dba(sampleSheet="NRF1_sample_sheet.csv")
```

After importing the data, it is possible to check the overall similarity of the samples based on peak location, by producing a heatmap showing pairwise Euclidean distances and the resulting hierarchical clustering of the samples (similar to the heatmap generated in Section 8.1).

The next step in `DiffBind` defines the consensus peak set that will be used for the comparison. The `minOverlap` parameter in the `dba.count()` function sets the requirement that a peak has to be present in the given number of samples to be included into the consensus set (default: 2). Repeating the heatmap visualisation of pairwise distances using the enrichment (input normalised read count) values at the consensus peak set will usually improve the clustering by sample type. `minOverlap` can be lowered to 1 in order to include all peaks, which might increase the noise, but decrease false negatives. The `dba.contrast()` function defines which column in the sample sheet is used for the comparison. Confounding parameters in the experimental setup can be taken into account by specifying their column in the `block` parameter. The `minMembers` parameter sets the minimum number of unique samples required in each compared group (default: 2). The final function to run the differential binding analysis is `dba.analyze()`. An important parameter to consider here is `bFullLibrarySize` deciding whether the size of the whole library is used for normalisation (default: TRUE). This is advisable for comparisons in which global changes are expected. Conversely, considering only read counts within peak regions (`bFullLibrarySize=FALSE`) is appropriate for comparisons for which less than half of the peaks are expected to change.

Note: Global changes might require normalisation using spike-in [48, 49].

```
# in R (~6 min)
# Set up and run the differential binding analysis
plot(NRF1)
NRF1=dba.count(NRF1, minOverlap=2)
NRF1=dba.contrast(NRF1, categories=DBA_CONDITION,
minMembers=2)
NRF1=dba.analyze(NRF1, method=c(DBA_DESEQ2))
```

DiffBind can export the differential peaks using dba.report() and visualise the results.

```
# in R
# Extract and visualise results
NRF1.DB=dba.report(NRF1)
dba.plotPCA(NRF1, contrast=1, label=DBA_CONDITION)
dba.plotMA(NRF1)
dba.plotBox(NRF1)
dba.plotHeatmap(NRF1, contrast=1, correlations=FALSE)
```

The PCA on differential peaks obtained from DiffBind again shows that both WT and TKO samples cluster closely between replicates and separate well between conditions. The MA plot shows the $\log_2$-transformed fold change of enrichment against the $\log_2$-transformed average enrichment for each peak in the analysis (Figure 8.2C). With the default cut-off at FDR<5%, DiffBind identifies 6,946 differentially bound peaks; most of them show stronger binding in the TKO cells. Note that this threshold is more lenient than the one in our DESeq2 analysis in Section 8.4.1, reflected in a higher number of identified changes. A boxplot shows the distribution of $\log_2$-transformed enrichment in all peaks compared to those that show significantly more (+) or less (-) binding in the WT cells (Figure 8.2D). In the case of the NRF1 data, we observe a strong shift in read density for all peaks, indicating that full library size normalisation is more appropriate for this dataset. Finally, the normalised enrichment per replicate for each differential peak can be visualised and clustered using a heatmap (Figure 8.2E). This again confirms that most peaks show higher enrichment in TKO, and that replicates have similar levels and therefore cluster together.

FURTHER READING

Bardet, AF, He, Q, Zeitlinger, J, Stark, A (2011). A computational pipeline for comparative ChIP-seq analyses. *Nat Protoc*, 7(1): 45–61.

Steinhauser S. et al. (2016). A comprehensive comparison of tools for differential ChIP-seq analysis. *Brief Bioinform*, 17(6): 953–966.

## CHAPTER 9

Downstream Analyses

Anaïs Bardet and Kathi Zarnack

I NTEGRATING ChIP-seq data with external information provides insights into the function of the protein of interest. This chapter introduces how to annotate the genomic context of the identified peaks and assign putative target genes that are then functionally characterised. It also suggests first steps to address the DNA sequence specificity of the studied protein and gives an outlook on how to integrate ChIP-seq with other functional genomics datasets.

9.1   GENOMIC CONTEXT

The genomic context of a transcription factor (TF) binding site can inform about its putative function in the cell. The genomic distribution of ChIP-seq peaks can be assessed at different levels of granularity, from global categorisation into chromatin types (colours) down to specific regions within individual genes.

9.1.1   Genomic location

In most cases, the first step is to examine the location of ChIP-seq peaks relative to annotated genes. However, the same approach can be applied to other genomic features, such as repetitive regions, CpG islands or enhancer regions.

Genes can be classified into protein-coding genes, pseudogenes and noncoding RNAs (referred to as gene biotype). Annotation comprises the transcribed region, but not the preceding promoter, which is commonly defined as 2 kb upstream of the transcription start site (TSS). The gene itself is divided into introns and exons, and further into 5' untranslated region (UTR), coding sequence (CDS) and 3' UTR in case of protein-coding genes. The remainder of the genome is referred to as intergenic.

Gene annotations (in GTF format) can be downloaded from ENSEMBL (ensembl.org), UCSC (genome.ucsc.edu) or NCBI (ncbi.nlm.nih.gov) as well as from species-specific resources (e.g. flybase.org, wormbase.org or arabidopsis.org). Depending on the source, annotation files can vary in terms of layout and information content. For instance, NCBI RefSeq annotation only includes a concise set of manually curated transcripts, whereas ENSEMBL reports the full spectrum of putative isoforms, including automatically annotated transcripts without experimental support. UCSC knownGenes is another widely used annotation source with a considerable number of transcripts. The most comprehensive annotation for long noncoding RNA (lncRNA) genes can be obtained from the PHANTOM project (fantom.gsc.riken.jp) [50].

Genomic features can also be retrieved through R/Bioconductor annotation packages (for more information see bioconductor.org/packages/devel/workflows/html/annotation.html). We provide alternative code in R for the genomic location analyses in the script available as Additional Online File.

Gene annotation can be difficult to work with since many features are overlapping. This occurs at the level of genes and is further amplified by multiple transcript isoforms for each gene. Overlapping annotations can be resolved by defining a hierarchy of annotations based on prior assumptions about the protein's function (e.g. exon > 5' UTR > 3' UTR > intron > promoter > intergenic) or by using an additional category (for example *ambiguous*). When using a hierarchy, caution needs to be taken to ensure that the resulting distribution is not due to the imposed hierarchy, and reflects the distribution of unambiguously assigned peaks.

The problem of overlapping annotations is further exacerbated by the fact that ChIP-seq peaks can be very broad. One way to resolve this is to only use TF peak summits. For broader regions, such as histone marks, the degree of overlap can be taken into account, such that e.g. >50% of the peak is required to lie within a given feature. Alternatively, regions can be assigned to multiple features by the fraction of overlap with each annotation.

The following code shows an example of how to assign the NRF1 and H3K27ac peaks to different genomic features. It uses

a pre-processed file for the mouse genome (mm10) based on the
ENSEMBL annotation of protein-coding genes (available as Additional Online File).

```
# Genomic location using the following hierarchy: exon
(EXON) > 5'UTR (5UTR) > 3'UTR (3UTR) > intron (INTRON) > 2
kb upstream promoter (P2000) > intergenic (INTER)
head mm10_genomic_features.bed

# Define TF peak summit positions
awk -v OFS="\t" '{print $1,$4-1,$4}'
peaks/NRF1_CHIP_WT_1_peaks_peakzilla.bed >
peaks/NRF1_CHIP_WT_1_peaks_peakzilla_summits.bed
# Genomic location of TF peak summits
intersectBed -wo -a
peaks/NRF1_CHIP_WT_1_peaks_peakzilla_summits.bed -b
mm10_genomic_features.bed | awk -v OFS="\t"
'{F[$7]++;t++}END{for(location in F){print
location,F[location]*100/t}}' >
peaks/NRF1_CHIP_WT_1_peaks_peakzilla_summits_location.txt
cat
peaks/NRF1_CHIP_WT_1_peaks_peakzilla_summits_location.txt

# Repartitioning of histone peak regions by genomic location
intersectBed -wo -a peaks/H3K27AC_CHIP_WT_1_peaks_macs.bed
-b mm10_genomic_features.bed | awk -v OFS="\t"
'{F[$10]+=$12;t+=$12}END{for(location in F){print
location,F[location]*100/t}}' >
peaks/H3K27AC_CHIP_WT_1_peaks_peakzilla_regions_location.txt
cat
peaks/H3K27AC_CHIP_WT_1_peaks_peakzilla_regions_location.txt

# Overall repartitioning of genomic location
awk -v OFS="\t" '{F[$4]+=$3-$2;t+=$3-$2}END{for(location in
F){print location,F[location]*100/t}}'
mm10_genomic_features.bed > peaks/genome_location.txt
cat peaks/genome_location.txt

# Plots
# Open R
R

# Load tables
tf = read.table("peaks/
NRF1_CHIP_WT_1_peaks_peakzilla_summits_location.txt")
histone = read.table("peaks/
```

```
H3K27AC_CHIP_WT_1_peaks_peakzilla_regions_location.txt")
genome = read.table("peaks/genome_location.txt")

# Merge the three tables
d = merge(merge(tf,histone,by=1),genome,by=1)
colnames(d) = c("Location","TF","Histone","Genome")

# Piecharts of the distribution of genomic locations
pdf("peaks/genomic_location_piechart.pdf",height=5,width=15)
par(mfrow=c(1,3),bg="white")
pie(d$TF,labels=d$Location,main="NRF1")
pie(d$Histone,labels=d$Location,main="H3K27AC")
pie(d$Genome,labels=d$Location,main="Genome")
dev.off()

# Barplots of enrichment over genome
pdf("peaks/genomic_location_barplot.pdf")
par(mfrow=c(1,2),bg="white")
barplot(log2(d$TF/d$Genome),names=d$Location,main="NRF1",
las=2,ylim=c(-2,6))
barplot(log2(d$Histone/d$Genome),names=d$Location,
main="H3K27AC",las=2,ylim=c(-2,6))
dev.off()

# Close R
q()
```

The results are shown as a piechart to represent the relative occurrence of peaks in different genomic regions (Figure 9.1A). Since the features occur with very different frequencies in the genome, the counts can be normalised by the total nucleotide count of each feature type in the genome, displayed as barcharts of enrichment (Figure 9.1B).

9.1.2 Distance to genes

TF binding sites can occur within promoter regions (proximal to TSS) or at intergenic locations (distal to TSS). In order to distinguish peaks that are located proximal or distal to TSS, the distance of each peak to the nearest TSS is examined, irrespective of a specific target gene assignment. Since many TFs bind both proximal and distal sites, the distances to TSS often show a bimodal distribution (Figure 9.1C). A different pattern is expected for histone marks which can also have positional preferences. For instance, trimethylation of lysine 27 on histone H3 (H3K27me3)

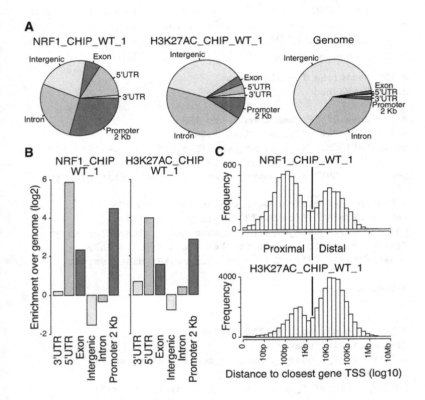

Figure 9.1 **Genomic context of peak regions.** (**A**) Piecharts of the distribution of NRF1 peak summits, H3K27ac peak regions and whole genome in different annotated genomic regions. (**B**) Barplots of the enrichment of the genomic distribution of NRF1 summits or H3K27ac regions over the genome. (**C**) Histograms of the distance of NRF1 and H3K27ac peaks to the TSS of the closest gene. Separation between proximal and distal regions is at 2 kb.

occurs almost exclusively in promoters, while the balance between H3K4me3 and H3K4me1 allows discrimination of promoters and enhancer regions [51].

The following code calculates each peak's distance to the nearest TSS. It uses a pre-processed file for the mouse genome (mm10) based on the ENSEMBL annotation of all transcripts of protein-coding genes (available as Additional Online File).

```
# TSS positions for all genes
ls mm10_tss.bed

# Distance of TF peak summits to TSS
awk -v OFS="\t" '{print $1,$4-1,$4}'
peaks/NRF1_CHIP_WT_1_peaks_peakzilla.bed | closestBed -d -t
"first" -a stdin -b mm10_tss.bed | awk '{print $NF}' >
peaks/NRF1_CHIP_WT_1_peaks_peakzilla_summits_dist_tss.txt

# Distance of histone peak region centre to TSS
awk -v OFS="\t" '{c=($2+$3)/2;print $1,c-1,c}'
peaks/H3K27AC_CHIP_WT_1_peaks_macs.bed | closestBed -d -t
"first" -a stdin -b mm10_tss.bed | awk '{print $NF}' >
peaks/H3K27AC_CHIP_WT_1_peaks_macs_center_dist_tss.txt

# Open R
R

# Load the distance to TSS tables
tf=read.table("peaks/
NRF1_CHIP_WT_1_peaks_peakzilla_summits_dist_tss.txt")
histone=read.table("peaks/
H3K27AC_CHIP_WT_1_peaks_macs_center_dist_tss.txt")

# Histograms of distances of peaks to closest TSS
pdf("peaks/dist_tss_hist.pdf")
par(mfrow=c(2,1),bg="white")
hist(log10(tf[,1]),main="NRF1",xlab="Distance to closet
gene TSS (log10)",breaks=seq(0,7,0.2))
hist(log10(histone[,1]),main="H3K27AC",xlab="Distance to
closet gene TSS (log10)",breaks=seq(0,7,0.2))
dev.off()

# Close R
q()
```

The results are shown as histograms displayed on a log_{10} scale to cover a broad range of distances (Figure 9.1C). This allows better visualisation of the bimodal distribution of proximal and distal peaks that characterises both NRF1 and H3K27ac binding. A natural separation of proximal and distal regions appears around 2 kb. NRF1 peaks are slightly more enriched proximal to genes representing binding in CpG-rich promoter regions, whereas distal binding sites have lower CpG content [9]. H3K27ac peaks are more enriched distal to genes representing the expected preference for distal active enhancer regions.

9.2 FUNCTIONAL ANALYSES

A popular downstream analysis is to explore the function of putative target genes. This relies on explicit peak-to-gene assignment.

9.2.1 Assignment to target genes

Peak-to-gene assignment remains a non-trivial task, since TFs and enhancers can activate their target genes from very long distances (e.g. the *sonic hedgehog (shh)* gene in mice is regulated by an enhancer located 1 Mb away [52]). Even though several concepts have been explored to assign target genes, the easiest and most efficient approach is to use the closest TSS [53]. Ideally, recently developed techniques, such as Capture Hi-C (CHi-C) [54], can be used to infer reliable associations, but data availability and processing remain limiting.

```
# Unique list of genes closest to NRF1 peaks (7167 peaks
leading to 5595 genes)
closestBed -t "first" -a
peaks/NRF1_CHIP_WT_1_peaks_peakzilla.bed -b mm10_tss.bed |
awk '{print $10}' | sort -u >
peaks/NRF1_CHIP_WT_1_peaks_peakzilla_genes.txt
```

9.2.2 Gene ontology analysis

A comprehensive description of gene functions is available across many species in the form of Gene Ontology (GO) annotation (geneontology.org) [55]. GO is organised into three non-overlapping ontologies, which describe a protein's physiological role (Biological Process), molecular activity (Molecular Function) or position within the cell (Cellular Component). Moreover, each GO term assigned to a protein is associated with an evidence code, specifying whether the assigned function was e.g. validated experimentally or merely inferred from orthology.

Based on GO annotations, a list of genes can be tested for enrichment of specific functions. For each GO term, the fraction of genes from the list that are associated with this term is compared to its overall occurrence to identify terms that are significantly over-represented. Significance is often calculated using a p-value from a hypergeometric test. It is important to note that GO enrichment can be strongly influenced by the choice of the baseline, i.e. whether enrichment is tested against all genes in the genome or a specific set of control genes. Commonly applied control sets are all expressed genes (e.g. according to RNA-seq data) or genes with shared versus differential ChIP-seq peaks. Popular online tools for GO enrichment analysis include DAVID (`david.ncifcrf.gov/`) as well as REViGO for visualisation of results (`revigo.irb.hr/`).

> Note: Similar to the choice of thresholds used for peak calling (see Chapter 6.2.5), enriched GO categories should always be ranked and selected based on p-values and not fold-changes. An arbitrary selection of GO terms should be avoided when reporting or visualising GO analysis results. The complete table of enriched GO terms should be made available as supplemental information.

9.2.3   Other gene-set enrichment analyses

The concept of enrichment can be extended to any predefined list of genes that is of interest in the context of the study. For instance, the target genes can be tested for an enrichment of developmentally regulated genes or interaction partners of a certain protein. The reference lists can be retrieved from publications or databases or compiled manually.

Another popular source of functional annotation is the KEGG database (`genome.jp/kegg/`) which collects manually curated biological pathways. Originally designed for enzymes and metabolic processes, KEGG now contains hundreds of manually drawn maps, including human diseases and drug design [56]. The KEGG Mapper tool allows to map a gene list onto the pathway maps which can be coloured according to the user-defined information. Finally, tools like g:Profiler (`biit.cs.ut.ee/gprofiler/`) integrate a broad range of different functional annotations into one conjoint resource to enable a comprehensive functional interpretation of gene lists.

9.3 SEQUENCE ANALYSIS

Analysing the DNA sequences underlying peak regions provides insights into the DNA binding preference of the studied protein or potential co-factors that recurrently bind at neighbouring positions.

9.3.1 Motif analysis

De novo **motif discovery** The first strategy in motif analysis is to search without *a priori* assumptions for sequences enriched in peak regions, also called *de novo* motif discovery. The search is usually performed in a window of 50-200 bp around the TF peak summits or whole regions for histone marks. Most motif discovery tools follow either a word-based or profile-based approach [57]. In the word-based approaches, implemented e.g. in DREME [58], all possible k-mers (i.e. sequences of length k) are exhaustively enumerated to generate consensus motifs that occur with increased frequency in the input sequences. In contrast, profile-based approaches, such as MEME [59], iteratively optimise sequence alignments to obtain the best scoring motifs. More recently, deep learning methods are applied to uncover binding motifs in ChIP-seq data [60].

Motifs are present with a high frequency across the genome. Therefore, any enriched motif should always be tested against control sequences, either provided by the user or generated by randomisation. The choice of these control sequences may strongly influence the discovered motifs.

HOMER is a popular tool that can be run from the command line (see example code below). It takes the genomic coordinates of both target and control regions as input or generates random control regions with the possibility to match the GC content of the target regions. MEME-ChIP [61] is a so-called ensemble tool which combines several motif discovery algorithms. It can be run as an online tool that takes FASTA sequences for both target and control regions as input or uses a random control following background letter frequencies.

Note: Motifs are represented as position weight matrices (PWMs) that are built from a multiple sequence alignment. A PWM reports the probability of occurrence of each nucleotide for each position in the motif, which can be visualised as a sequence logo.

```
sample=NRF1_CHIP_WT_1

# Define regions of 151 bp around peak summits
awk -v OFS="\t" '{print $1,$4-75,$4+75}'
peaks/${sample}_peaks_peakzilla.bed >
peaks/${sample}_peaks_peakzilla_151bp.bed

# HOMER for de novo motif search (~1 h)
findMotifsGenome.pl
peaks/${sample}_peaks_peakzilla_151bp.bed genomes/mm10/
motifs -size given

# Open the html page with results
see motifs/homerResults.html &
```

HOMER outputs a ranked list of motifs found in the target sequences (Figure 9.2A). For each motif, it indicates the sequence (represented in a logo), a p-value corresponding to the enrichment of this motif in target compared to control sequences as well as the best match for this motif within known motifs. In the case of NRF1, a *de novo* identified motif matching the known NRF1 motif is found as most enriched in the peak regions, as expected. It is present in 64% of our target peak regions.

Known motif search A second strategy in motif analysis is to scan the peak regions for already defined motifs, also called known motif search. Motifs for many TFs have now been derived from *in vitro* (e.g. using systematic evolution of ligands by exponential enrichment (SELEX) [62] or protein binding arrays (PBM)) or *in vivo* (e.g. using ChIP-seq) experiments and are available in public databases (e.g. JASPAR [63] or HOCOMOCO [64]). PWMs of known motifs can be used to scan the genomic regions of interest to identify motif occurrences (e.g. using MAST [65]). To select significant motif occurrences, a p-value threshold needs to be applied, which we suggest to adjust according to the information content of the motif (i.e. the same threshold will have a different stringency depending on the length of the motif). The following code shows how to search for the known NRF1 motif in our peak regions.

```
# Download the NRF1 motif from JASPAR
# jaspar.genereg.net/matrix/MA0506.1
wget http://jaspar.genereg.net/api/v1/matrix/MA0506.1.meme
-O motifs/NRF1.meme
```

```
# Scan genome for motif occurrences using a p-value
threshold of 10^-5 and reformat output into BED (~5 min)
mast -hit_list -mt 1e-04 motifs/NRF1.meme genomes/mm10.fa |
awk '($1!~/#/){if($2=="+1"){s="+"}else{s="-"};print
$1,$3-1,$4,"NRF1",$6,s}' | gzip > motifs/NRF1_mm10.bed.gz

# Number of peak regions: 7167
cat peaks/${sample}_peaks_peakzilla_151bp.bed | wc -l
# Number of peak regions with NRF1 motifs: 5245 (73%)
intersectBed -u -sorted -a
peaks/${sample}_peaks_peakzilla_151bp.bed -b
motifs/NRF1_mm10.bed.gz | wc -l
```

Using the known NRF1 motif at a specific threshold, we find
that 73% of our target peak regions harbour a motif. In ChIP-
seq data for TFs, the proportion of peaks with motif is usually
around 60-80%. Some non-specific peaks could arise from experimental biases, such as crosslinking artefacts. The same code could
be adapted to run on control regions (e.g. generated using the command **shuffledBed**). Alternatively, sub-selections of peaks could
be used, such as TKO-specific versus shared peaks. Finally, the
comparison of enrichment in target versus control regions can be
assessed statistically using a hypergeometric test (e.g. using the R
command **phyper**). The same analyses can be run for many more
motifs or even all possible k-mers. Compared to the *de novo* motif
discovery approach, the advantage of using known motifs to scan
the peak regions is that this information can be used for further
analyses such as exploring the organisation and co-occurrence of
different motifs in specific regions (e.g. distance to each other or
orientation). Moreover, calculating positional enrichment in metaplots allows us to visualise if and where the motifs are enriched
around peak summits.

9.3.2 Sequence conservation

When multiple alignments with additional species are available,
the peaks' or motifs' conservation level can be explored. To do
so, conservation scores such as PhastCons [66] or PhyloP [67] are
available for download in a bigWig format from the UCSC genome
browser and can be processed with **bwtool** or **bedtools** (see Section 9.4.1).

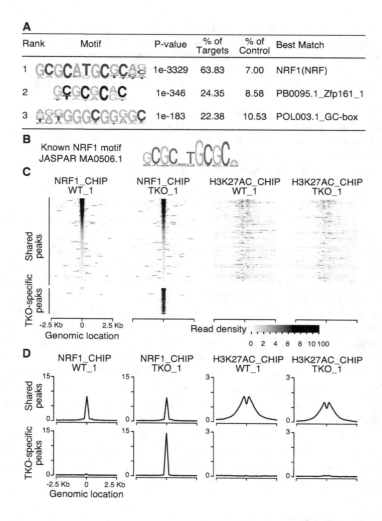

Figure 9.2 **Integration with motifs and other ChIP-seq datasets.**(**A**) First three motifs found by *de novo* motif search using HOMER. (**B**) Known NRF1 motif logo from the JASPAR database. (**C**) Heatmaps of NRF1 and H3K27ac read densities in WT and TKO samples for shared and TKO-specific peaks. Data is centred around NRF1 peak summits and ranked according to NRF1 peak scores from replicate 1. (**D**) Metaplots summarising the heatmap signal from (C).

9.4 INTEGRATION WITH OTHER DATASETS

Genomic studies often entail several types of experiments to address a specific biological question. Additionally, a profusion of related genomic datasets can be publicly available. Therefore, integration of the ChIP-seq data with other data types is a common analysis. An example of such data integration can be found in the original publication of the NRF1 datasets [9].

9.4.1 Additional ChIP-seq datasets

The first step is often the integration with other ChIP-seq datasets, which could include a combination of data for TFs and histone marks.

Note: In order to avoid any bias and mis-interpretation, it is strongly recommended to process each type of datasets (or re-process public data) using a similar pipeline including data pre-processing (e.g. read length, trimming), read mapping (e.g. index, filtering threshold for unique reads) and peak calling (e.g. algorithm, peak threshold).

A popular approach to visualise and compare several ChIP-seq datasets for TF and histone marks is to generate a heatmap of the read densities in peak regions. This integration should take into account the different nature of the identified peak regions: the signal of a histone mark is usually broader and peaks around the TF signal. Therefore, it is recommended to perform the comparative analysis on regions centred on specific positions, such as TF peak summits or TSS rather than merging all enriched regions. Below, we provide code to generate such heatmaps for NRF1 shared and differential peak regions across samples. There are several user-friendly tools available online to generate heatmaps and other exploratory plots from sequencing data (e.g. deepTools2 [12]).

```
# For each peak selection, for all samples, extract read
densities (from the bigwig file) for each position within 5
kb around the peak regions (~2 min)
for peaks in TKO_spec WT_spec shared
do
    for sample in NRF1_CHIP_WT_1 NRF1_CHIP_TKO_1
    H3K27AC_CHIP_WT_1 H3K27AC_CHIP_TKO_1
    do
        awk -v OFS="\t" '{center=int(($2+$3)/2);print
```

```
      $1,center-2500,center+2500,$2,$3}'
      changes/NRF1_all_regions_count_${peaks}_table.txt |
      bwtool extract -tabs bed stdin tracks/${sample}.bw
      stdout >
      changes/NRF1_all_regions_count_${peaks}_density_$
      {sample}.txt
   done
done

# Open R and load library
R
library(gplots)

# Load RPKM table
rpkm=read.table("changes/NRF1_all_regions_rpkm.txt")
colnames(rpkm)=c("chr","start","end","NRF1_CHIP_WT_1",
"NRF1_CHIP_WT_2","NRF1_CHIP_TKO_1","NRF1_CHIP_TKO_2")

# For each peak selection
for(peaks in c("TKO_spec","shared")){
    png(paste("changes/NRF1_all_regions_count_",
    peaks,"_density_heatmap.png",sep=""))
    par(bg="white",mfrow=c(1,4))

    # For each sample
    for(sample in c("NRF1_CHIP_WT_1","NRF1_CHIP_TKO_1",
    "H3K27AC_CHIP_WT_1","H3K27AC_CHIP_TKO_1")){

        # Load the density table
        x=read.table(paste("changes/NRF1_all_regions_count_",
        peaks,"_density_",sample,".txt",sep=""))

        # Add extra column with matching RPKM of region in
        NRF1_CHIP_WT_1 sample
        y=merge(x,rpkm[,c("chr","start","end",
        "NRF1_CHIP_WT_1")],by.x=c(1,4,5),
        by.y=c("chr","start","end"))

        # Plot image of read densities, ordering regions
        (row) by RPKM and adjusting colour scale from white
        to black by 0.1 steps between 0 and 10 and one more
        step until 100
        image(t(y[order(y$"NRF1_CHIP_WT_1"),7:5006]),axes=F,
        col=colorpanel(101,"white","black"),
        breaks=c(seq(0,10,0.1),100),main=sample)
    }
```

```
    dev.off()
}

# Plot colour scale
pdf("changes/
NRF1_all_regions_count_density_heatmap_scale.pdf",height=3)
par(bg="white")
plot(c(0,110),c(0,1),xlab="Read
density",ylab="",pch="",axes=F)
axis(1,at=seq(0,110,10),labels=c(0:10,100))
cols=c(colorpanel(100,"white","black"),rep("black",10))
rect(seq(0,109,1),0,seq(1,110,1),1,border=cols,col=cols)
dev.off()

# Close R
q()
```

The density heatmaps display the distribution of read densities around peak regions in shared and TKO-specific peaks (Figure 9.2C). It could also be generated from the whole list of peak regions from Chapter 8.3.

Note: It is important to keep in mind that although heatmaps are nice visualisation tools, they are not an exact way to represent data since subtle variations in the colour scale can be misleading for the human eye. In the example of the density heatmaps generated here, since thousands of regions are displayed in a relatively small figure, if the rows would not be ordered by descending RPKM values, some regions with low read densities would not be visible between regions with high read densities. Additionally, it is easy for the user to arrange the colour scale in non-linear steps to highlight specific characteristics e.g. in our case, we used a linear scale from white to black from 0 to 10 and annotated all additional values above 10 until 100 to black since the density values follow a descending exponential curve.

9.4.2 Expression data

Integrating ChIP-seq with gene expression information, e.g. from RNA-seq or microarray data, allows us to monitor whether the binding of a TF or the presence of a histone mark relates to the expression of its target genes. To do so, the signal in the peak regions can be compared to the expression level of the putative target genes. If several conditions are available, it can be more informative to compare changes in binding and gene expression in a so called delta-delta scatterplot (see Chapter 8.3.4). Note that since several peaks can be assigned to the same gene, some gene expression values might be represented multiple times. This could be resolved e.g. by taking the minimum, maximum or average signal of all associated peaks for a given target gene.

> Note: This analysis is sensitive to the noise introduced by false-positive target genes when assigned to the closest TSS of a gene. Since distal peaks might be more often wrongly assigned than proximal peaks, it can be useful to separately perform the downstream analyses for proximal and distal peaks (e.g. $\leq$ 2 kb vs. > 2 kb).

9.4.3   Other types of data

Finally, other types of genomic data could also be integrated into the analysis such as chromatin accessibility (e.g. DNase-seq or ATAC-seq) or DNA methylation (e.g. whole-genome or reduced representation bisulfite sequencing). This can be performed by summarising information over peak regions using `bwtool` or `bedtools` (see Section 9.4.1). For example, signal averages can be calculated over peak regions or compared in relevant subsets of peak regions. Similarly to comparisons with gene expression, delta-delta scatterplots can be used to compare changes in binding with changes in chromatin accessibility or DNA methylation.

In addition to being used for target-gene assignment, data from high resolution Hi-C-based methods can also be integrated by comparing changes in ChIP-seq binding with changes in interaction profiles of the genomic regions containing the differential peaks.

FURTHER READING

Blake, J.A. (2013). Ten quick tips for using the gene ontology. *PLoS Comput Biol*, 9(11): e1003343.

du Plessis, L., Skunca, N., and Dessimoz, C. (2011). The what, where, how and why of gene ontology–a primer for bioinformaticians. *Brief Bioinform*, 12(6): 723–735.

Lihu, A., and Holban, Ş (2015) A review of ensemble methods for de novo motif discovery in ChIP-Seq data. *Brief Bioinform*, 16(6): 964–973.

Bibliography

[1] A., R.J., Spacek, D.V. *et al.* (2015). High-throughput sequencing technologies. *Mol Cell*, 58(4):586–597.

[2] Rhee, H.S. and Pugh, B.F. (2012). ChIP-exo method for identifying genomic location of DNA-binding proteins with near-single-nucleotide accuracy. *Curr Protoc Mol Biol*, Chapter 21:Unit 21.24.

[3] He, Q., Johnston, J. *et al.* (2015). ChIP-nexus enables improved detection of in vivo transcription factor binding footprints. *Nat Biotechnol*, 33(4):395–401.

[4] Skene, P.J. and Henikoff, S. (2017). An efficient targeted nuclease strategy for high-resolution mapping of DNA binding sites. *eLife*, 6:pii: e21856.

[5] van Steensel, B. and Henikoff, S. (2000). Identification of in vivo DNA targets of chromatin proteins using tethered dam methyltransferase. *Nat Biotechnol*, 18(4):424–428.

[6] Chen, Y., Negre, N. *et al.* (2012). Systematic evaluation of factors influencing ChIP-seq fidelity. *Nat Methods*, 9(6):609–614.

[7] Egelhofer, T.A., Minoda, A. *et al.* (2011). An assessment of histone-modification antibody quality. *Nat Struct Mol Biol*, 18(1):91–93.

[8] Kharchenko, P.V., Tolstorukov, M.Y. *et al.* (2008). Design and analysis of ChIP-seq experiments for DNA-binding proteins. *Nat Biotechnol*, 26(12):1351–1359.

[9] Domcke, S., Bardet, A.F. *et al.* (2015). Competition between DNA methylation and transcription factors determines binding of NRF1. *Nature*, 528(7583):575–579.

[10] Afgan, E., Baker, D. *et al.* (2018). The Galaxy platform for accessible, reproducible and collaborative biomedical analyses: 2018 update. *Nucleic Acids Res,* 46(W1):W537–W544.

[11] Kallio, M.A., Tuimala, J.T. *et al.* (2011). Chipster: user-friendly analysis software for microarray and other high-throughput data. *BMC Genomics,* 12:507.

[12] Ramirez, F., Ryan, D.P. *et al.* (2016). deepTools2: a next generation web server for deep-sequencing data analysis. *Nucleic Acids Res,* 44(W1):W160–165.

[13] Cock, P.J., Fields, C.J. *et al.* (2010). The Sanger FASTQ file format for sequences with quality scores, and the Solexa/Illumina FASTQ variants. *Nucleic Acids Res,* 38(6):1767–1771.

[14] Patel, R.K. and Jain, M. (2012). NGS QC Toolkit: a toolkit for quality control of next generation sequencing data. *PLoS ONE,* 7(2):e30619.

[15] Dodt, M., Roehr, J.T. *et al.* (2012). FLEXBAR-Flexible Barcode and Adapter Processing for Next-Generation Sequencing Platforms. *Biology (Basel),* 1(3):895–905.

[16] Bolger, A.M., Lohse, M. *et al.* (2014). Trimmomatic: a flexible trimmer for Illumina sequence data. *Bioinformatics,* 30(15):2114–2120.

[17] Ye, H., Meehan, J. *et al.* (2015). Alignment of Short Reads: A Crucial Step for Application of Next-Generation Sequencing Data in Precision Medicine. *Pharmaceutics,* 7(4):523–541.

[18] Langmead, B. and Salzberg, S.L. (2012). Fast gapped-read alignment with Bowtie 2. *Nat Methods,* 9(4):357–359.

[19] Li, H. and Durbin, R. (2010). Fast and accurate long-read alignment with Burrows-Wheeler transform. *Bioinformatics,* 26(5):589–595.

[20] Weirather, J.L., de Cesare, M. *et al.* (2017). Comprehensive comparison of Pacific Biosciences and Oxford Nanopore Technologies and their applications to transcriptome analysis. *F1000Res,* 6:100.

[21] Treangen, T.J. and Salzberg, S.L. (2011). Repetitive DNA and next-generation sequencing: computational challenges and solutions. *Nat Rev Genet,* 13(1):36–46.

[22] He, X., Cicek, A.E. *et al.* (2015). De novo ChIP-seq analysis. *Genome Biol*, 16:205.

[23] Chen, J., Rozowsky, J. *et al.* (2016). A uniform survey of allele-specific binding and expression over 1000-Genomes-Project individuals. *Nat Commun*, 7:11101.

[24] Li, H., Handsaker, B. *et al.* (2009). The Sequence Alignment/Map format and SAMtools. *Bioinformatics*, 25(16):2078–2079.

[25] Carroll, T.S., Liang, Z. *et al.* (2014). Impact of artifact removal on ChIP quality metrics in ChIP-seq and ChIP-exo data. *Front Genet*, 5:75.

[26] Planet, E., Attolini, C.S. *et al.* (2012). htSeqTools: high-throughput sequencing quality control, processing and visualization in R. *Bioinformatics*, 28(4):589–590.

[27] Landt, S.G., Marinov, G.K. *et al.* (2012). ChIP-seq guidelines and practices of the ENCODE and modENCODE consortia. *Genome Res*, 22(9):1813–1831.

[28] Bailey, T., Krajewski, P. *et al.* (2013). Practical guidelines for the comprehensive analysis of ChIP-seq data. *PLoS Comput Biol*, 9(11):e1003326.

[29] Noble, W.S. (2009). How does multiple testing correction work? *Nat Biotechnol*, 27(12):1135–1137.

[30] Wilbanks, E.G. and Facciotti, M.T. (2010). Evaluation of algorithm performance in ChIP-seq peak detection. *PLoS ONE*, 5(7):e11471.

[31] Zhang, Y., Liu, T. *et al.* (2008). Model-based analysis of ChIP-Seq (MACS). *Genome Biol*, 9(9):R137.

[32] Bardet, A.F., Steinmann, J. *et al.* (2013). Identification of transcription factor binding sites from ChIP-seq data at high resolution. *Bioinformatics*, 29(21):2705–2713.

[33] Johnson, D.S., Mortazavi, A. *et al.* (2007). Genome-wide mapping of in vivo protein-DNA interactions. *Science*, 316(5830):1497–1502.

[34] Gotea, V., Visel, A. *et al.* (2010). Homotypic clusters of transcription factor binding sites are a key component of human promoters and enhancers. *Genome Res*, 20(5):565–577.

[35] Berman, B.P., Nibu, Y. *et al.* (2002). Exploiting transcription factor binding site clustering to identify cis-regulatory modules involved in pattern formation in the Drosophila genome. *Proc Natl Acad Sci USA*, 99(2):757–762.

[36] Rhee, H.S. and Pugh, B.F. (2011). Comprehensive genome-wide protein-DNA interactions detected at single-nucleotide resolution. *Cell*, 147(6):1408–1419.

[37] Heinz, S., Benner, C. *et al.* (2010). Simple combinations of lineage-determining transcription factors prime cis-regulatory elements required for macrophage and B cell identities. *Mol Cell*, 38(4):576–589.

[38] Ibrahim, M.M., Lacadie, S.A. *et al.* (2015). JAMM: a peak finder for joint analysis of NGS replicates. *Bioinformatics*, 31(1):48–55.

[39] Guo, Y., Mahony, S. *et al.* (2012). High resolution genome wide binding event finding and motif discovery reveals transcription factor spatial binding constraints. *PLoS Comput Biol*, 8(8):e1002638.

[40] Kent, W.J., Sugnet, C.W. *et al.* (2002). The human genome browser at UCSC. *Genome Res*, 12(6):996–1006.

[41] Robinson, J.T., Thorvaldsdottir, H. *et al.* (2011). Integrative genomics viewer. *Nat Biotechnol*, 29(1):24–26.

[42] Bardet, A.F., He, Q. *et al.* (2011). A computational pipeline for comparative ChIP-seq analyses. *Nat Protoc*, 7(1):45–61.

[43] He, Q., Bardet, A.F. *et al.* (2011). High conservation of transcription factor binding and evidence for combinatorial regulation across six Drosophila species. *Nat Genet*, 43(5):414–420.

[44] Li, Q., Brown, J.B. *et al.* (2011). Measuring reproducibility of high-throughput experiments. *Ann Appl Stat*, 5(3):1752–1779.

[45] Tu, S. and Shao, Z. (2017). An introduction to computational tools for differential binding analysis with chip-seq data. *Quantitative Biology*, 5(3):226–235.

[46] Love, M.I., Huber, W. *et al.* (2014). Moderated estimation of fold change and dispersion for RNA-seq data with DESeq2. *Genome Biol*, 15(12):550.

[47] Robinson, M.D., McCarthy, D.J. *et al.* (2010). edgeR: a Bioconductor package for differential expression analysis of digital gene expression data. *Bioinformatics*, 26(1):139–140.

[48] Bonhoure, N., Bounova, G. *et al.* (2014). Quantifying ChIP-seq data: a spiking method providing an internal reference for sample-to-sample normalization. *Genome Res*, 24(7):1157–1168.

[49] Egan, B., Yuan, C.C. *et al.* (2016). An Alternative Approach to ChIP-Seq Normalization Enables Detection of Genome-Wide Changes in Histone H3 Lysine 27 Trimethylation upon EZH2 Inhibition. *PLoS ONE*, 11(11):e0166438.

[50] Hon, C.C., Ramilowski, J.A. *et al.* (2017). An atlas of human long non-coding RNAs with accurate 5' ends. *Nature*, 543(7644):199–204.

[51] Heintzman, N.D., Stuart, R.K. *et al.* (2007). Distinct and predictive chromatin signatures of transcriptional promoters and enhancers in the human genome. *Nat Genet*, 39(3):311–318.

[52] Lettice, L.A., Heaney, S.J. *et al.* (2003). A long-range Shh enhancer regulates expression in the developing limb and fin and is associated with preaxial polydactyly. *Hum Mol Genet*, 12(14):1725–1735.

[53] Sikora-Wohlfeld, W., Ackermann, M. *et al.* (2013). Assessing computational methods for transcription factor target gene identification based on ChIP-seq data. *PLoS Comput Biol*, 9(11):e1003342.

[54] Mifsud, B., Tavares-Cadete, F. *et al.* (2015). Mapping long-range promoter contacts in human cells with high-resolution capture Hi-C. *Nat Genet*, 47(6):598–606.

[55] Ashburner, M., Ball, C.A. *et al.* (2000). Gene ontology: tool for the unification of biology. The Gene Ontology Consortium. *Nat Genet*, 25(1):25–29.

[56] Kanehisa, M., Furumichi, M. *et al.* (2017). KEGG: new perspectives on genomes, pathways, diseases and drugs. *Nucleic Acids Res*, 45(D1):D353–D361.

[57] Lihu, A. and Holban, S. (2015). A review of ensemble methods for de novo motif discovery in ChIP-Seq data. *Brief Bioinformatics*, 16(6):964–973.

[58] Bailey, T.L. (2011). DREME: motif discovery in transcription factor ChIP-seq data. *Bioinformatics*, 27(12):1653–1659.

[59] Bailey, T.L. and Elkan, C. (1994). Fitting a mixture model by expectation maximization to discover motifs in biopolymers. *Proc Int Conf Intell Syst Mol Biol*, 2:28–36.

[60] Alipanahi, B., Delong, A. *et al.* (2015). Predicting the sequence specificities of DNA- and RNA-binding proteins by deep learning. *Nat Biotechnol*, 33(8):831–838.

[61] Ma, W., Noble, W.S. *et al.* (2014). Motif-based analysis of large nucleotide data sets using MEME-ChIP. *Nat Protoc*, 9(6):1428–1450.

[62] Jolma, A., Yan, J. *et al.* (2013). DNA-binding specificities of human transcription factors. *Cell*, 152(1-2):327–339.

[63] Khan, A., Fornes, O. *et al.* (2018). JASPAR 2018: update of the open-access database of transcription factor binding profiles and its web framework. *Nucleic Acids Res*, 46(D1):D260–D266.

[64] Kulakovskiy, I.V., Vorontsov, I.E. *et al.* (2018). HOCO-MOCO: towards a complete collection of transcription factor binding models for human and mouse via large-scale ChIP-Seq analysis. *Nucleic Acids Res*, 46(D1):D252–D259.

[65] Bailey, T.L. and Gribskov, M. (1998). Combining evidence using p-values: application to sequence homology searches. *Bioinformatics*, 14(1):48–54.

[66] Siepel, A., Bejerano, G. *et al.* (2005). Evolutionarily conserved elements in vertebrate, insect, worm, and yeast genomes. *Genome Res*, 15(8):1034–1050.

[67] Pollard, K.S., Hubisz, M.J. *et al.* (2010). Detection of nonneutral substitution rates on mammalian phylogenies. *Genome Res*, 20(1):110–121.

Index

Printed in the United States
by Baker & Taylor Publisher Services

Printed in the United States
by Baker & Taylor Publisher Services